Beyond The Toxic Harvest

A chemical poisoning survivor story-
Beating the odds, making miracles

Eve Hillary

Beyond The Toxic Harvest

Health House

PO BOX 5165, Chittaway Bay, NSW, Australia, 2261
sarahslastwish@gmail.com

ISBN 978-0-9806629-3-1

Non-Fiction: Biography, health, medical, environmentally induced illness, chemical injury, pesticide poisoning, integrative medicine, and complementary medicine, spiritual.

Acknowledgments

I wish to thank those people who have given me encouragement, love, friendship, inspiration, outright help and the courage to write this book. Special thanks go to Beverley M, Beverley and Edwin K, Gina S, Ted K, Chris W, Mary N, Christopher P, Fabia V, Pamela V, Lois and Derek B, Ruth H, Colleen S, Rhana S, the Westley family, and my professional colleagues at the neurosurgical intensive care unit at the Royal North Shore Hospital. Since this book's first writing, some have passed on, but they will live on in my heart and in this story.

My eternal love and thanks go to my sons for bringing love and joy into my life, and for providing me with the best reasons to overcome all that stands in the way of living my best life. I thank them for putting up with me while I was trying to learn the art of living.

Thank you to my husband who is an inspiring example of what can be achieved in a lifetime.

My love and devotion remains with my remarkable parents, William and Anne, my sister Karin and her children.

Thanks to Dr Kate Short for having the courage to write the first book in Australia about the effects of pesticide exposure on humans, many years ago, when the topic was still taboo. Thanks to Peter Harding for starting a support group for people with chemical injury. Thanks to Bill Henty for his friendship, legal skills and good judgment and a special thanks to my editor Beverley M.

I will always be indebted to Dr Gunnar Heuser, Dr. Emmanuel Varipatis and Dr Mark Donohoe for helping us when no one else would or could. I thank these extraordinary professionals for showing their patients compassion and for having the courage to diagnose and treat those with the politically incorrect disease of chemical injury.

Thanks to my loyal Cattle Dogs, Emma and Suzi for their tireless service, love and loyalty. And to Suzi, who was there with me every minute while writing this book.

Finally, I dedicate this book to all who have suffered from chemical injury and received little or no understanding and help, and to the memory of those who have taken their own lives because their suffering has been too great — may they rest in peace.

Author's Note:

Pseudonyms have been used, where necessary, including for boats, to protect the identity of some people, and in a few cases personal details have been changed. These changes, however, do not detract from the essential truth of the story.

Foreword

I found myself reading Eve's story on many levels: as a story of her life-journey and challenges, as an exploration of the issue of chemical injury and its effects on people, and as a challenge to look at the broader implications of this problem.

The ability of human beings, like Eve, to transcend suffering, and somehow to find themselves, their strength, direction and spirituality through this same suffering, is one of life's miracles. That we also reject anything or anyone who is different, who threatens our view of reality is, however, one of the limitations that we must strive to overcome.

So, what is chemical injury? Doctors such as myself are seeing people like Eve and her son every day, in ever-increasing numbers. They range from mildly affected people who feel generally unwell in urban environments, to those who will suffer major physical symptoms and illness for days after being exposed to the chemicals in consumer products. Once someone's system is overloaded and damaged, then the chemical doses required to cause severe illness become infinitesimal. This is chemical injury. If these people are indeed the 'canaries' of our society, then heaven help us all in fifty years.

By writing a book like this, Eve may help comfort the millions of people with chemical injury who feel rejected and ridiculed by society and by their doctors. It will help increase public awareness of this problem, and hopefully will hasten the day when we start making the hard decisions necessary to protect our world, our children and our futures.

Dr Emmanuel Varipatis

Table of Contents

Hopefully Ever After

On a day

When the wind is perfect,

The sail just needs to open and the world is full of beauty.

Today is such a day.

-Rumi

I could not have asked for better weather on my wedding day in October 1981. The fresh nor'easter is Sydney's trademark wind, and it blew in another spring day that was perfect for brides and sailors.

Sydney, Australia boasts the highest number of small boats per capita of any city in the world. This Friday lunchtime was no different from any other, when pleasure sailors emerged from their high rise offices in Sperry Topsider deck shoes, ready to board their boats, and slip the marina for open water. Like a plague of Rock Lobsters, half the city's rich, famous and successful headed into every Sydney cove to cleat on a mooring for the weekend and enjoy a few drinks.

Bill and I would soon join them, but first we were due at the registry office to get married. Only a few months before, Bill had been in a sentimental mood one afternoon when he half knelt down in front of me and mumbled something about applying for our passports. When I nodded, he suggested we go on a honeymoon to the US and Europe. My mouth was still open when he added: 'I suppose we should get married before we go on a honeymoon'.

Bill's style won me over.

At half past noon, Bill and I stood before the civil celebrant in the functional grey room of the Chippendale registry office. The solitary spring posy in my hands marked the solemn occasion of our marriage. Bill looked dignified in his navy suit coat and grey pants. He was 59 years old and still in his prime with a sturdy body and a noble European face. On his feet he wore wing tip lace up shoes; it was the first and only time I saw him without deck shoes.

Bill was never one for open displays of affection, but he looked at me lovingly that day, while I stood awkwardly beside him in my silk pant-suit, blushing like a teenager.

I was 29 years old and thought I knew about life. I had a solid career as an RN behind me with working experience in a number of major teaching hospitals. My second career came about by accident after a professional photographer took some photos of me. They were good enough to start me off on a career as a runway and photographic model at Joan McInnes's model agency. These two polar opposite vocations left me with a model's legacy of never feeling attractive enough, and a nurse's legacy of thinking about everyone else's comfort, except my own. I needn't have worried though, the only other people there were two of Bill's friends who were quite capable of looking after themselves, as well as our flowers and wedding rings.

The room was otherwise empty; neither Bill nor I had much family, at least none in Australia. Mine lived in the US and his were in Europe and the UK. None of our relatives had been given enough notice to come over for the wedding. Bill had preferred to arrange it quickly and wanted to keep it simple. And simple it was; even my 7 year old son, Ben, was at school that day instead of at the wedding, an arrangement that I have regretted ever since. However, Bill was an alpha male, used to calling the shots, and I didn't challenge him much in those early days. He had a forceful and dynamic personality, and expected to marry a traditional wife. I might have been one of the last women in the Western world - after Germaine Greer published the Female Eunuch - willing to sign up to a traditional marriage, but I loved and admired him enormously, and was willing to compromise.

The celebrant, a straight laced bureaucrat, began the marriage ceremony from behind a desk as we deferentially stood before her like two school children in the head master's office. She asked us to formally declare that we were over 18. This raised a chuckle from my husband to be. The celebrant smiled indulgently and continued, asking us next to declare that we were free and not legally married to someone else. I could feel Bill's arm tensing as he no doubt thought about his late wife, Sally, who had died of a heart attack the year before.

I felt myself blush too, as emotional memories stirred. I had worked to help my first husband though his last two years of medical school. My parents had bought us a building block and also helped us out financially over those two years. Unfortunately, after graduation he left the marriage for another woman, and moved to Western Australia. That left me to shoulder the full responsibility of raising our three year old son without his support. After staying in Adelaide for a few more difficult years as a single parent, I was determined to give Ben the best chance that he could have in life, and decided to pack up and move us to Sydney, where I started to feel free again to pursue a new life.

I heard Bill respond to the celebrant's question and responded in turn, but I knew that we both still had emotional scars from our past, and neither of us knew how to become free of them, except to start over.

But a higher purpose was at work in bringing us together. Bill and I also had a lot of common ground: we were both migrants, torn from our original cultures. He had escaped a totalitarian regime in Europe in 1939 and arrived in Australia, broke and hopeful, as a 19 year old refugee. And I arrived as a teenager with my parents, when they migrated here in 1969. A few years later they returned to the US and left me here to make my own life in Australia, without any other member of my family here. Bill and I both knew what it felt like to be transplanted and to practically perish before being assimilated into a new culture. Our resilience shaped us both into doers who thought that after the confronting the challenges we had survived, we could achieve almost anything.

By the 1960s Bill had become hugely successful in business and was a well known yachtsman who had won several prestigious international ocean races throughout the 1960s and 70's. Through his achievements, he attained fame, a prominent place in society, and dined with princes and prime ministers. I admired him for the drive and determination it takes to create that kind of success without family support.

I was also sure that we were attracted to each other, in part, because we wanted to recreate the family that neither of us had. I saw our union as an opportunity to build together as a team, and to create a wonderful life for Ben and our future children, (that included Bill's grown son and step daughter). Bill and I agreed to try for a child as soon as possible, before he was much older. Like so many brides, I hoped for a utopian future where his, mine and ours would blend into a large extended family.

The celebrant asked me: 'do you take this man to be your lawful husband?'

'Yes', I said, surprised at my own shyness.

'Do you take this woman to be your lawful wife?' she asked Bill.

'I do', he answered.

'You may exchange rings'.

Bill's friend searched around in his pocket and stepped forward with my wedding band, which Bill awkwardly slipped on my finger.

'I pronounce you husband and wife', said the celebrant.

From the rear of the room, a photographer appeared and directed us to pose for pictures; one of us holding hands, one kissing and a more formal shot for the mantelpiece. Meanwhile, the celebrant had placed the marriage certificate on her desk where she invited us to sign it, and asked Bill's friends to witness our signatures. The date on the certificate was October 15, 1981. After two more photos, the formalities were over and we were married.

An hour later, Bill and I were cruising Sydney Harbour in a classic World War II Navy Pinnace with an assortment of Bill's friends and a case of Vintage Veuve Clicquot on board. The women guests wore the windswept hairdo of Princess Diana, who had been married to Prince Charles only a few months earlier; while the men shared war stories about past yacht races and wholeheartedly quaffed the bubbly.

I knew few of the guests, since I'd arrived in Sydney only 14 months before, but I was in Bill's world now, and happy to get to know his friends, since they would feature in our lives for many years to come.

By four that afternoon we docked at the Cruising Yacht Club marina in Rushcutter's Bay. After seeing off the last of the wedding guests, we drove home to Bayview, on Sydney's North Shore, to the house that Bill had built in the late 1950s. It was small, but comfortable, and nestled in bush land on the high side of a leafy street, with a spectacular view overlooking Pittwater Bay.

As a married couple, we came home to unite our new family, and I couldn't wait to see Ben. I had made arrangements for him to stay with the next door neighbors, but I knew he would, instead, be at his friend Peter's house and race home as usual, just before dark. When he appeared, I gave him a big hug and told him about the wonderful things the future held for all of us.

We spent the next few months as three happy people who had just received everything they had always wished for. Bill and I had found each other, Ben enjoyed family life again and the future seemed guaranteed to bring health, family and good times.

Bill even got to indulge his travel bug on our honeymoon. In November we returned from visiting our families in the UK, Europe and the US, and just when we thought life couldn't get better, the year topped out with our first Christmas together.

Eve with Ben around the time of her marriage to Bill

Frontier Honeymoon

And I thought of old things, and I thought of old folks,
Till I sighed in my heart to the sigh of the oaks;
For the years waste away like the waters that leak
Through the pebbles and sand of
Eurunderee Creek.

- Henry Lawson

I could not have imagined that straight after the New Year of 1982 - just when I was enjoying sitting on the patio at Bayview and looking out over Pittwater and Lion Island - I would become a farmer's wife on Bill's outback sheep and cattle property. Bill declared that from now on, we would spend most of our time living on the grazing property near Mudgee, on the Western Tablelands, and return to Bayview only on long weekends and holidays.

Our first trip out west seemed endless. Ben's car sickness flared up around the winding mountain roads. Even I wondered why anyone would want to regularly take a four hour trip through the deep gorges and high plateaus of the Great

Dividing Range. But Bill broke up the trip halfway at the coal mining town of Lithgow where we stopped for a Shandy, a roast beef sandwich and a spectacular view of the Hartley Valley Gorge.

Just before dusk we arrived at the front gate of the property that Bill had aptly named 'Landfall' - a nautical term that represents the safely and comfort of land after a long sea voyage.

Bill asked me to get out of the car and open the front gate, and then close it again after he had driven through the entrance. He explained that the homestead was only another two km down the property's private road. As I stood holding the gate open, I looked out over the rolling green hills and valleys and sensed a strange feeling that this land was primordial and timeless. After closing the gate and getting back into the car, Ben and I gazed through the car window in awed silence as Bill drove us toward our new home. The property opened out onto breathtaking, undulating granite country — 5068 acres (2051 hectares) of rolling hills, weathered granite escarpments, green valleys, creek flats and cool dams.

We arrived at the hundred-year-old homestead, just in time to see its corrugated iron roof and stone chimney reflect the last glint of orange sunset. At first the old home looked lonely with its brooding windows facing the paddock dam where a rickety windmill creaked cheerlessly, but Bill was determined to make a good first impression. He parked the car in the driveway and rushed out to open the homestead door for us, while Ben and I lingered a while, before going inside, to savor the pungent fragrance that wafted from the nearby trees. A row of peppercorn trees, heavy with ripening peppercorns, overhung the drive, and from that moment, I associated the spicy sweet perfume with the bush house.

Inside the homestead, Bill had turned on the lights and waited in the doorway for us. Ben and I carted our bags through the entrance and glanced around the roomy living and dining area. From the living room I ventured into the narrow kitchen and peered through the window to discover a view to the driveway and up the hill opposite the house where another cottage stood. Bill explained that this cottage was once used as accommodation for the various farmhands or managers that occasionally lived there. 'Let's go on a cook's tour', said Bill, as he raced ahead to show us the rest of the house. He led us through the dining room with its wicker furniture cozily arranged and facing a row of glass windows that looked out over a hedge.

'It's too dark now', said Bill, 'But in the morning the roses are really sensational - no thanks to me'.

I realized that the garden reminded Bill of his late wife, Sally. Like Bill, I had no green thumbs, but made a mental note to keep the roses looking as good as possible, to honor Bill's memory of her.

From there, Bill led us through a brick edged doorway into the winter room, which was in the dead centre of the house, without windows to the outside, but its charming rustic brick fireplace made up for its darkness.

Behind the TV, in a corner, hung a burlap curtain that separated the winter room from the infrequently used rear section of the house. Ben was intrigued and swept the curtain aside, revealing a gloomy passageway.

'Where does this go?' asked Ben.

Bill explained that it led into an old disused part of the house, which had been derelict for years. 'No one lives there anymore. At least no one living', said Bill as he led us away from the area and toward the bedrooms. Ben and I glanced at each other but didn't ask about it further for now. Instead, I carted my suitcase into the master bedroom where huge sunflower print wallpaper covered the walls, while Ben settled into his bedroom across the hall from ours.

The homestead was a typical bush dwelling consisting of a labyrinth of hallways, leading into a patchwork of rooms, all added on at different times, over the years. I was sure that Ben would explore every nook over time, but for now, I was ready for dinner and bed after the long day. Tomorrow promised even more excitement when Bill planned to show us around the property.

At dawn the next morning, I woke to find myself alone in bed with the sun streaming through the sheer curtains, lighting up the yellow sunflower wallpaper in a blinding display of optimism. After a quick shower, I hustled Ben out of bed.

Ben and I found Bill already seated at the dining room table enjoying the same breakfast that he'd eaten each morning, without fail, since I'd known him; two half grapefruits, toast with marmalade and a cup of black coffee. He wouldn't allow us to rush his ritual and we just had to wait impatiently until he had finished, before he was willing to tackle starting his Land Rover. Bill explained that since the Land Rover was one of the earlier models from the 1960s, it often had a flat battery, and this morning was no different; it had to be started with a crank handle.

When the old Rover spluttered to life, Bill's dog, a speckled brown German Shorthair Pointer named Polly, jumped in the back. Bill drove us past the house dam and windmill to the first paddock gate. It was Ben's job to jump out of the Rover, open the gate, wait for Bill to drive through and close the gate behind us, before taking a running leap back into the Land Rover when Bill took off again. The first paddock gate was also Polly's cue to leap off the back and start her marathon run alongside the Land Rover. The dog had endless stamina and thought nothing of running 30 km, almost on a daily basis, whenever Bill checked the property.

With Polly racing alongside the vehicle, Bill first headed toward the eastern side of the property where many of the paddocks shared a border fence with the Munghorn Gap Nature Reserve, a national park that was magically filled with colourful native birds. These pasturelands were on higher elevations and offered some of the most spectacular views of the thousands of acres of Landfall's lush hills and valleys. The area had been inhabited, since pre-history, by the Wiradjuri Australian Aboriginal people and was often shrouded in mist and steeped in an air of mystery. Bill stopped on the summit of

Mount Molly to let us out and take in the view. When I first looked over Bill's property from its highest elevation, the vista stretched out as far as I could see and I thought I'd died and gone to heaven. It was another world, where time stood still.

From there, Bill headed north east, into the heart of Landfall where he drove us through 40 paddocks, each one around 125 acres in size, with gentle hills, valleys and fertile creek flats. Scattered around the flats were three hundred Hereford cattle with calves that stood knee deep in clover. By ten in the morning they'd already had enough of the rich fodder, and were lying in groups under shady gum trees, where they hardly noticed us driving by.

In the higher paddocks, where the shorter native grasses grew, were scattered 10 000 grazing sheep with thick yellow fleeces. Most ewes had lambs at foot; many with twins and some with some triplets. I had seen nothing like it before and could hardly take it all in.

It took us until midday to reach the rear of the property, and just when I thought it couldn't get any better, we arrived in the back paddock, which was one and a half times the size of the country of Monaco, where twenty Arab horses had the run of the 605 acre (245-hectare) spread. This turned out to be a pasture breeding arrangement, seen rarely these days, where the mares all ran in a herd and bred naturally with the stallion. Until then, I had only seen mares and geldings in small paddocks or stables – with stallions locked up separately. I was thrilled and fascinated to see horses in their natural herd setting with the foals frolicking within sight of the grazing mares.

Watching over the herd from a grassy rise was a white pure-bred Arabian stallion, Shimaal, who flared his nostrils and turned his ears in our direction when we approached in the Rover and parked among them. In a display of raw masculine power, the stallion lifted his tail straight up, pawed the ground and pranced on the spot until the entire group of mares joined him in a fluid and graceful dance around the Land Rover.

When the lead mare stopped the parade and faced our direction expectantly, it was Bill's cue to reach for the oats bucket from the back of the vehicle and offer some oats to her. The alpha mare had claimed the privilege of eating first, and the rest waited respectfully for her to begin. I watched the mare's muscled jaw disappear into Bill's oat bucket and noticed Ben standing behind Bill, hoping to pat the horses. Unfortunately, as soon as the mare had her first bite, the remaining herd descended on the bucket in an alarming free-for-all. Bill was holding his own and waving the horses off, but I was horrified to see my seven-year-old had disappeared in the middle of twenty horses that were biting and pushing each other to get to the feed bucket.

Just when I thought Ben had been trampled to death, I saw him emerging from the skirmish, dusty but very much alive and running for cover toward me and the safely of the vehicle. Meanwhile, Bill dumped the oats on the ground and headed back himself.

After giving Ben a reassuring hug, I told Bill that I was annoyed with him for putting Ben in harm's way, but he only laughed and dared me to take Ben's place next time.

'That's crazy'. I said. 'Well, maybe when the stallion is out of the way', I added grumpily.

As a teenager, I had dreamed of having a horse, but my wildest dreams couldn't have conjured up the idea of having a whole herd of them. I had been around racehorses in Thoroughbred stables, but not an entire pack of unbridled Arabians. I wondered how they could be handled and educated; how the young ones would be weaned or trained, and whether some of them were broken in and could be ridden. Most of all, I wondered how anyone could handle any of them fresh out of a herd with a stallion clearly in charge. Those answers would come, through hands-on experience, sooner than I could imagine.

We arrived back home at sunset. After Polly had run all day she slumped contentedly onto the front door mat. But before going inside, Bill drove us up to the fuel shed, located on a small elevation above the homestead and directly across from the workmen's cottage. There, he showed us where the fuel drums were, in case I had to fuel up any of the four farm vehicles from the 44 gallon drums.

It was pitch dark inside the shed, but for a weak shaft of light from the late afternoon sun that illuminated a cluster of 44 gallon drums standing on the oily floor. While Bill showed me how to connect the hand pump to the diesel drum, my nose and eyes became irritated by the stinging chemical fumes coming from a row of drums lined up against the wall. These, said Bill, contained pesticides and herbicides. When Bill had finished showing me the ropes, I was glad to be out of there. Stinging eyes should have been a warning of things to come, but it didn't dampen my enthusiasm for life on the land.

While preparing dinner that night, I decided that this was the life for me. Ben was thrilled with his new home too, judging by the grin on his face. The sun had tanned and freckled his face, and I thought the fresh country air was already doing him good. When I was his age, I'd spent two years on my grandparents' farm. It turned out to be the highlight of my childhood and I was glad that Ben, too, could experience life close to plants, animals and Nature.

Over dinner that evening I told my husband about my happy and carefree lifestyle on my grandmother's farm. She had been widowed in her early 40s but carried on raising steers, pigs, chickens and rabbits, and she grew all her own fruits and vegetables. She never used farm chemicals of any kind and her plants and animals were always healthy, while her pastures were fertile after using compost, animal manure and crushed rock minerals on them. Grandmother was the original biodynamic farmer, even before the word was invented. She always impressed on me that people who lived in harmony with nature would always have everything provided and never needed to worry, but those who went against nature would pay the high price of becoming sick and unbalanced.

Bill listened to my story, but I could see that after a while his patience wore thin.

'This is very different,' he said sternly, adding that I should forget about my granny's 'Mickey Mouse' outfit and come into the twentieth century, where agribusiness is now organized and scientific. The set-up had to show an annual profit and a yearly increase, just like any other company.

'My dear', said Bill finally, 'this is not a hobby farm'.

'I suppose so', I said, since I was hardly in a position to argue with Bill about agriculture, when he owned and managed a sheep and cattle property more than twice the size of the country of Gibraltar.

The following week I enrolled Ben at the local primary school at Mudgee, an hour's round trip from our property. Every morning I drove Ben 6 km to the bus-stop on Ulan Rd, where he caught the school bus and passed through some of the most historic areas of New South Wales on his way to school, including Cook's Gap, Frog Rock, Eurunderee and Pipe Clay Creek, the early home of Henry Lawson, the Australian bush poet.

During the Colonial era of the mid 1800s, the Mudgee area had attracted hordes of gold prospectors; among them migrants from America and Europe who made primitive camps along the creeks to pan for gold. Henry Lawson's parents were among them. His Norwegian father met his Australian mother at the Eurunderee Creek mining camp and young Lawson spent his early childhood years there. Despite the rough and difficult conditions, those early days inspired some of Lawson's most sentimental poetry.

I sensed a rich history in the region and used every excuse to drive Ben in to school myself, so I could duck into the town's library afterwards. But it was more than just history that intrigued me. Since coming to the area, I'd wondered what secrets the land held about the Wiradjuri Indigenous people who inhabited the land before Europeans came and I also wanted to read more Henry Lawson, hoping the bush poet's words could reach across time to reveal what life was like in the Colonial days after Europeans arrived in the region.

Warning Signs

For whatever happens to the beasts

Also happens to man. All

Things are connected.

— Chief Seattle,
Suqwamish and Duwamish tribe

The warm February nights under the pungent peppercorn trees had now turned into a cold June winter that enveloped Landfall in heavy morning frosts. 1982 was one of the coldest winters on record and it brought a rare and spectacular event.

Ben had settled well into Mudgee Primary school and he even managed to enjoy the daily bus ride with some newfound friends that lived in the district. One morning we woke up to a fresh snowfall that blanketed the landscape. Since I grew up in Michigan, USA, I was used to snow, but I had never seen it in Australia. It was the first time Ben had seen snow in his life, and he was beside himself with excitement.

On the way to the bus-stop we encountered one snow drift after another, until Ben excitedly demanded I stop the car to let him out. For the next few minutes I watched my son frolic in the snow and make snowballs until his hands were red from cold. It could have been a scene from the wilds of Michigan until a half dozen big red kangaroos came bounding over the hill. After jumping a 1.5 m (5 ft) fence, they started breakfasting on the tender shoots and leaves of native shrubs right in front of us. That morning Ben had a lot to share with his class for show and tell.

Each morning I returned from the bus stop, put on my gumboots and turned up for the regular morning meeting that Bill had with the stockmen and farm hands.

Bill's manager, Harry, and his son Warren were full time employees, while Bill hired contractors for seasonal jobs, such as shearing.

Harry and Warren turned up each morning in a red utility and sauntered over to where Bill was waiting for them in the driveway outside the homestead. Harry was a typical Australian country-bred man with a weather-beaten face, flat backside and bowed spine, while nineteen year old Warren was a young, still unbent, version of Harry. As much as they looked alike however, both had different personalities. Harry was a slow, deliberate mover who never unhooked his thumb from his rawhide belt and never said anything unless he had something important to say. In contrast, young Warren still had the bravado of youth and could smooth talk a couple of local girls while steering the utility with one hand and shooting a rabbit with a .22 rifle using the other. Warren was also young enough and skilled enough to be one of the tough breed of Australian stockmen who'd worked in the Northern Territory and learnt to break in wild Brumbies for outback stations.

In my very first week I witnessed Warren's horsemanship first hand when he cut a young colt out of the horse herd in the 640 acre back paddock and drove him, on horseback, square across the property to the horse yards near the homestead. From a safe distance, I watched speechless as he lassoed the skittish unhandled yearling in the lunging yard, and had him quiet and tied to a post, with a halter on, in under an hour.

Each year, Bill let Warren have the pick of the yearlings from the rear paddock. They were bred out of our stock mares and sired by our pure bred Arab stallion, Shimaal, and Warren turned them into good stock horses, and sold them for a tidy profit. Warren's horse trading with our Arabian stock meant that our bloodlines had infused a lot of blue blood into the district's stock horses over the years.

Once Warren had the colt under control, I came in for a closer look at the dark, bluish grey two-year-old that stood, still sweating, after being broken in. The colt had the dished face and regal bearing of his Arab sire, Shimaal, while his mother, a Thoroughbred stock mare, had given him a fine set of long legs.

'I'm calling this one Blue Boy', Warren announced as he grinned broadly at the young horse. No doubt he was already thinking about the lordly image he would cut astride the Anglo-Arabian colt at the next camp-drafting competition, where he was a legend in the making among the horse wranglers and young lovelies in the district.

The first week of my 'apprenticeship' on Landfall was entirely devoted to horses. Warren and Harry mustered in the thundering herd from the back paddock and drove them into the front stock yards to be wormed, vaccinated and to have the farrier trim their hooves. This was dangerous work and not for anyone worried about getting horse-kicked.

I watched from a safe distance and took every opportunity to retreat inside the homestead to study the pedigrees of the purebred Arabians and match up the paperwork with the corresponding horses in the paddock. Horse breeding had been the exclusive domain of Bill's late wife, Sally, and Bill was only too happy to pass on the responsibility to me, since he was completely uninterested in horses. 'The bloody things eat me out of house and home', he would grumble.

Without the slightest bit of help from Bill, I figured out that a smaller group of mares and their foals were kept in the front paddock, between the homestead and the front gate, away from the herd at the rear of the property. They were expensive purebred Arabian mares, sired by the legendary Arabian stallion Ralvon Pilgrim. Sally had been lucky enough to secure these valuable pedigrees in her purebred foundation broodmares, which she sent interstate once a year to be bred to an imported Arabian stallion. The mares had foaled magnificent purebred Arab fillies last season and they needed to be bred again this year after the fillies were weaned. But before I could make any decisions about selecting a stallion for them, I needed to trace the pedigrees back ten years in order to discover where Sally was headed with the breeding program. It was a lengthy search and a bit of a guessing game, but eventually I got the general idea and started to match up suitable stallions to our purebred mares for the next season. I never got much praise from Bill for sorting out the horse stud, but I loved the challenge anyway, and being around such a legendary lineage of Arabian horses was its own reward.

Each afternoon, I knocked off farm work around 4.30 pm to pick Ben up from the bus stop on the main road. He was always eager to show me his schoolwork, and some of his English assignments demonstrated a budding literary mind. Ben was becoming quite the bush poet himself, as though Henry Lawson had channeled his poetry into my young son each time he passed by the poet's childhood home.

'Mum, what's at the back of the house?, asked Ben one day, on our way home.

'Don't know, you'll have to ask Bill', I said.

Once home, Ben raided the fridge as he usually did after school. While he fixed himself some crackers and cheese, I searched for Bill, since I wanted to ask him the question that was on Ben's mind.

I found Bill lighting the fireplace in the winter room. 'What's at the back of this house?' I asked.

He turned and looked surprised. 'You mean you haven't been in that old part of the house yet?' he asked.

I shook my head.

'It's the old homestead, before I added this section on', Bill explained. 'The bloke who lived here before I bought the place got sick of living during the 1966 drought'.

My eyes widened. 'You mean he…'

'Shot himself'. Bill finished my sentence.

I looked toward the doorway to see if Ben was within hearing distance. 'Do me a favor Bill', I pleaded in a whisper. 'Please don't tell Ben that'.

Just then, Ben came in, balancing his chocolate Milo and plate of crackers. He asked Bill the same question.

'It's just the original homestead before I added this bit on', explained Bill.

I don't know if Ben heard what Bill had said to me earlier, but after he'd finished his snack he called Polly and took her with him exploring, as he usually did, around the nearby hills and dams.

Like most boys in the bush, Ben enjoyed making things from the odds and ends he found around the hay-sheds and tool-sheds. I never had to entertain him with TV or electronic gadgets; he was so interested in exploring his environment that he lost track of time. I had to call him when it got dark, and was always relieved to see him and Polly bounding home for dinner. I needn't have worried though; Polly protected Ben from both worldly and otherworldly threats. The only place he and the dog consistently avoided was the abandoned old section at the rear of the homestead.

On cold winter evenings, Bill lit the fireplace and we ate dinner on trays in the warmth of the winter room. Bill insisted on steak every night, while Ben and I were happy with less red meat and more vegetables. After Ben went to bed, Bill relaxed on his old Parker lounge with a rum and cola, and gave dissertations. He maintained that agribusiness has come such a long way that we could now control nature with technology.

'How so?' I asked.

Bill relished a teaching opportunity. 'I can seed a pasture and coax it up with double super phosphate fertilizer and carry much more stock than nature ever intended', he explained.

'What about weeds and insects?' I asked.

'Easy', he said. 'I can make one phone call and eradicate pests and weeds with the sweep of a crop duster'.

'What about water?'

'Good question', praised Bill. 'Over the last couple of years, I've bulldozed five new dams and expanded old ones, so there won't ever be a problem watering stock in the dry spells'.

The next day Bill took me with him in the Land Rover to show me what he was talking about and pointed out what kind of special pasture grasses were needed to fatten livestock quickly. We made this into a daily ritual and I soon understood how Bill managed this huge agricultural holding. If he found a problem, it wasn't unusual to see Harry on the tractor the next day, reseeding the paddock with clover or ryegrass to keep the property pasture improved so it could carry more stock. The following day a plane would fly overhead and aerially drop super-phosphate fertilizer.

If Bill found weeds or pests on our inspection tours, the plane turned up again, this time with herbicides or pesticides raining down from its wing sprayers. It seemed that anything unwelcome had to be rooted out to keep the production line going.

There was so much Bill expected me to learn, and he counted on me to understand what he meant.

I still knew nothing about planting fodder crops such as oats, barley and Lucerne, or about fencing and shearing. When I tried asking Harry and Warren questions, they would tip their Akubras and say, 'Where 'ave youse been?' I learnt mainly by observation and imitation and didn't let it bother me. I knew the men enjoyed testing me to see if I would last the distance or retire to the homestead nursing my blisters and bruised ego. In those early days I had unlimited energy and, since I was the only woman on the place, I wasn't going to be beaten down by a couple of cocky stockmen.

September brought the annual calving season and also the second lambing of the year, and I was thrown into the deep end. Bill and I went around checking the cattle and sheep daily to see if any were calving problems, where we would sometimes need to assist in difficult deliveries. This usually posed no problem in case of a lamb, but pulling a calf sometimes required several people and the cow did not always survive the procedure.

Despite some stock losses, Bill was getting phenomenal calving and lambing numbers. Even in the 1980s, animal reproductive technology was so advanced that entire herds could be induced to ovulate at the same time with synthetic hormones, to ensure births occurred at an economically advantageous time. Although we did not use artificial reproductive technology, we did promote rapid growth after weaning by fattening young stock on our lush Lucerne, sorghum and oat paddocks. The end result was that the local meat buyers always bid hard for our animals at the sale-yards, since they were turned out well fattened with a healthy sheen. But a huge amount of work was needed before each animal reached that stage and our workload included: branding, ear tagging, tail docking lambs, castrating male calves and lambs, vaccinating young stock and looking after sick and injured animals.

In the middle of this intensive period of work I found my very own niche, one that my hospital background gave me a good grounding for: animal husbandry. Before too long I drove to the local feed store in town and purchased an array of veterinary supplies including: injectable antibiotics, vaccinations and various powders, tinctures and spray-on antiseptics for open wounds on horses, sheep and cattle. Bill was so chuffed about saving money on vet bills that he got as close as he ever came to openly praising me in front of the men.

By spring, thousands of sheep and cattle, along with their lambs and calves, needed treatments to make sure the parasite levels were kept down while the animal numbers were so high. This meant I had to learn more about the multitude of farm chemicals stored in our fuel shed.

Over the next several months, I ran around the paddocks herding the bleating mobs of sheep through a succession of gates while Bill had the easy job of heading them toward the stock yards from the relative comfort of the Land Rover. In the next paddock Harry and Warren were on horseback and lounged back in their stock saddles while the cattle dogs did the hard work. When we'd mustered a few hundred lambs and ewes into the yards, we put them through a race, one by one, to drench and dip them in small lots. Drenching was to kill worms and intestinal parasites; it involved wedging the sheep in a narrow race and squirting the anti-worm agent into its mouth with a drenching gun. Farmhands wore a drenching pack containing the liquid chemical on their back like a rucksack, with the packs often leaking chemical down the wearer's buttocks.

Dipping was entirely different and far more hazardous. It consists of a chemical application to the animal's skin to kill ticks or lice, and to prevent fly-strike, which is an infestation caused by fly larvae that feed on the living tissue of sheep, causing agonizing death to stock and huge economic losses to graziers. Until the 1950s, dipping was done using deadly arsenic compounds in a bath; often little more than a concrete-lined hole in the ground. When the government banned arsenic due to its toxicity, farmers switched to the organo-chlorine (OC) group of chemicals, including DDT, Lindane, Aldrin and Dieldrin. By that time, chemical showers had replaced the old dipping baths and sheep were herded past spray nozzles dispersing the chemical, which often ended up soaking the farmhands as well. The government banned the OC group of chemicals in the 1970s, to protect the export markets, since the chemicals left residues in the animals' tissues. After that, the government directed farmers to use the organo-phosphate (OP) family of compounds to treat sheep and other livestock.

By the 1980's, when I was involved with this kind of work, the dipping baths and showers were no longer in use. Instead, we painted organophosphate chemicals on areas of the animal's skin every 8 weeks, where it would permeate the hide or fleece and repel parasites for weeks before another application was needed. On most properties, the regular use of different types of toxic chemicals over many years caused stockyards to become highly toxic environments, a problem which we were about to discover.

One day Bill and I were helping the farmhands drench a herd of Herefords in the cattle yards while Polly was milling around the yards with the working dogs. Being a hunting breed, Polly had a good nose and enjoyed sniffing the interesting smells around the yards. Underneath one of the water tanks was a puddle that had formed when the farmhands had washed their drenching and dipping equipment under the tank's faucet. Bill's dog must have taken a drink from the puddle instead of the water bowl left out for the working dogs, since, when we looked up from our work, she was staggering around the yards drunkenly. Bill and I watched in disbelief as the dog became agitated and spun around in circles for a while, before arching her back and dropping into the dust unconscious and convulsing. We climbed over the stockyard railing and found Polly lying on her back, eyes staring vacantly, and frothing from the mouth. It was obvious we had to race her to the vet in town. While Bill rushed up to the homestead to get the family car, I checked the dog over. Her pupils seemed unusually constricted to pinpoints; a symptom I had only seen in accident and emergency when patients came in after overdosing on opiates. It was an unlikely scenario for our dog, but I couldn't think what else could have that effect. Soon Bill turned up in the car and we struggled to lift our convulsing dog into the back seat, where I slid in beside her. When I laid my hand on her chest I felt her heart beating so forcefully that I thought she would have a cardiac arrest on the way. Even though Bill had flattened the pedal to the floor, it was obvious that Polly was in her death throes and I wasn't sure if we would get there in time. I felt powerless to help her, as she was certainly well beyond comforting. Perhaps poor Polly was fortunate to be unconscious, since whatever was wringing the life out of her looked too painful to endure. I only hoped the vet would be available to deal with our emergency.

We made it to Mudgee in just under a half hour and it took both of us to carry the unconscious dog up the steps and into the surgery waiting room. The vet poked his head out of the consultation room and made the diagnosis without hesitation. He told the receptionist to show us in while he rushed back into his treatment room to draw up a vial into a syringe. We felt panicky as we lifted Polly onto the stainless steel examination table, and wondered why the vet seemed so matter of fact, but soon realized that he already knew what the problem was and stood waiting for us with a safety razor and an antidote.

The vet hastily shaved Polly's foreleg to reveal a vein. 'It's organophosphate poisoning', he said, as he slipped the needle in and plunged the antidote into her bloodstream. Within seconds, our dog stopped convulsing and in another minute, she opened her eyes groggily and regained consciousness.

'Atropine is the antidote to OPs', said the vet. 'You're lucky you came in time', he added, as he cleared away the syringe.

While Bill was enthusiastically patting his dog back to life, I thought about what just happened. I knew about atropine, since I had often administered it to hospital patients as premedications. I also knew that it blocked the effects of certain drugs on the central nervous system, but I was curious to know whether atropine blocks the action of organophosphate pesticides, and if these chemicals are nerve toxins.

The vet seemed pleased to answer my question. 'Organophosphates are potent cholinesterase inhibitors in animals', he said. 'Cholinesterase is an important enzyme needed for the nervous system to function. The chemicals that inhibit this enzyme act as potent neurotoxins. Cholinesterase inhibitors are in sheep-dips and drenches'.

That was interesting information, but what he left out proved to be an even more important discovery that I made some years later: cholinesterase inhibitors are so potent that they are found in some snake venoms and in deadly nerve gasses used in war. Worst of all, exposures to OP's cause birth defects and neurological damage in humans.

On our way home, Polly remained dazed, but at least she was alive. I remember thinking that I should have some OP antidote on hand, just in case another dog gets a dose of poison, and I made a mental note to add atropine to my vet supply shopping list. I also felt in awe of science for saving our dog's life and, like Bill, I believed that science was a tool that could be used to control the course of life and death. My education in healthcare was typically Western scientific reductionism, which compartmentalized and artificially separated humans, animals and plants from the ecosystem as a whole. I assumed (as did the vet) that a chemical that harmed animals was not necessarily a threat to humans, and that the government would not allow a product on the market unless it was safe for humans when used as directed.

Despite the prevailing lax attitudes about chemicals, Polly's poisoning alerted me to be more careful with the chemicals around people. Between now and the end of the year we had only a small group of 50 merino sheep to drench and dip. Before Bill and I pitched in to help the farmhands, I was careful to read the directions on the tin of sheep dip. I discovered the active ingredient was diazinon, and it came with a tame warning: 'Use as directed, wash skin after using, irrigate the eyes if necessary, and if ingested, seek medical advice immediately'.

We had no problems mixing up the concentrated product as directed, in a portion of water, just as we had done the previous times, before applying it to the remaining 50 sheep. I took that as proof of the product's safety and thought Polly's poisoning was just an unfortunate fluke. But, in future I would keep Polly out of the yards while we worked.

It was soon the end of November and apart from Polly's mishap, we had spent a busy year on the property, relatively free of major problems.

Bill was watching the news one night when a report came on about a dry spell in South Australia and Victoria. The newsreader added that so far, the New South Wales tablelands were unaffected.

'Let's pack up and go to Bayview for the Christmas break', announced Bill.

'But what about Polly?' I asked. 'I mean, so soon after that freak accident'.

'The boys'll look after her. She'll be alright'.

Normally Harry and Warren took care of Polly when we were at Bayview, but I didn't think she'd regained her fitness after the incident and might fret while we were away. On the other hand, I really needed some time out at Bayview, and we could always phone Harry to see how everything was going on the property.

Two weeks later, Ben came home from school clutching his end-of-year school photo. It showed a healthy, tanned and alert looking boy with his broad smile showing his obvious happiness about the end of the school year and the coming summer holidays. He'd finished grade two at the Mudgee Primary School with a wholehearted appreciation for Australian Bush poetry, while gaining precious knowledge about life on the land, which gave him special memories for the rest of his life. In his schoolbag, I was thrilled to find some of his original poetry.

THE DARLEY RACE

The horses in the starting gates
Were ready for the race
A jockey named Sir Johnny Yates
Was thought to be the ace
But a fella from old Mudgee town
Was out to prove 'em wrong
And as he rode Blue Thunder home
He began to sing his song:
'No Bugger from old Sydney town 'll
Beat me on the course -
'Cos I'm the bloke from Mudgee town
And Blue Thunder is my horse.'

- *Ben's Poetry*

On a hot and dry December day, we packed up Bill's roomy diesel Mercedes to head to Bayview for the holiday break. I had done alright to keep up with the farmhands during this challenging year, but there was still was no sign of a baby. I had lived in a pair of black gumboots for nearly a year and had almost forgotten what it was like to wear a pair of strappy sandals. Now I looked forward to feeling like a woman again and going shopping at Mona Vale for a dress and a pair of shoes.

Nature Strikes Back

In the beginning of all things,
Wisdom and knowledge were with the animals,
For the Creator did not speak directly to man, but showed
Himself through animals, so that Man
Should learn through them about
The Creator.

- Eagle Chief (Letakos-Lesa) Pawnee

Time out at Bayview over Christmas hit the spot for all of us, but for Ben, it was also a homecoming. Young boys have radar, and within minutes Peter and Jason wheeled round on their bikes and welcomed Ben back from his year long absence in the bush, as though he hadn't been gone at all. Peter, who lived at the end of the road, was a frequent visitor. Together the boys discovered Dick Smith Electronics and soon assembled tiny electronic eavesdropping devices to spy on their parents. I wasn't sure if I should be pleased about their inventiveness or annoyed about the invasion of privacy. I found myself sweeping the house for possible 'bugs' and hoping the boys would soon lose interest and move on to different hobby. Years later, I heard that Peter's mother did the same.

After stocking the pantry with holiday food and shopping for a new summer outfit and a pair of sandals, I looked forward to relaxing with Bill, only to discover that he was not much of a relaxer. While I stuffed the turkey and prepared the trimmings, Bill was on the phone to Harry, asking if it had rained in Mudgee. While we opened our Christmas presents, Bill surfed the TV channels, searching for weather reports. After Christmas lunch, Ben took off to Peter's house, I did the dishes, and Bill started on his calculator, working out stock numbers, beef and wool prices.

The next day, on Boxing Day, Bill was glued to the TV for the start of the Sydney Hobart Yacht Race. He had completed 15 of these classic yacht races when he'd owned his last series of boats, and his trophy cabinet was full of medals from competing in these and other major ocean races. But since spending the past three years 'boat-less', Bill felt like his sailing mates were leaving for Hobart without him. As cranky as Bill was on Boxing Day, I could always count on him to drive down to Mona Vale for any errand, knowing that he would have to pass by several boat marinas on the way. Marinas always meant stopping by for a chat with other yachties, and he would return from the simplest errand, hours later.

Just before the New Year, it was obvious that our attempts to have a child were still not successful, despite the rest from farm work. While we were close to the city's medical facilities, it seemed a good time to have some fertility tests done. When my tests turned out okay, it was time for Bill to be tested, but he was not keen on the idea of giving a 'sample' without some considerable encouragement. He reminded me that in the bush it was usual practice to send old bulls to the knackery when they no longer produced offspring. 'What if I don't pass the test? He asked mournfully.

But all was well in the end, Bill passed his test and we rang in the New Year of 1983 with new optimism.

Two weeks later, on a warm January day, I was on the patio enjoying the view over Pittwater, when Bill read me an article from the Land Newspaper. The wet season had failed in tropical North Queensland, an almost unheard of event.

'That's a bad sign. We have to go back to the bush tomorrow', insisted Bill.

I thought that wasn't fair on Ben who didn't have to be back at school for another two weeks and was enjoying his friends, but once Bill had made a decision, no one could change his mind, and sure enough, we were on the road the next day.

The closer we got to the Western Tablelands, the drier it got, with noting but scorched brown pastures for miles.

Driving through the front gate at Landfall, we noticed a different atmosphere than the one we had left only a few short weeks ago. The pastures had lost some of their green colour, the water level in the billabongs had dropped and the roads were dusty.

'This doesn't look good at all', said Bill as he drove us toward the homestead. At that stage we didn't realize how close we were to a one in a thousand year weather phenomenon. The south eastern wheat crop had failed and most of Victoria had no rain since November. Creeping toward us, on the Western Tablelands of New South Wales, was a super drought.

When we arrived at the homestead, Ben rushed out to greet Polly. She was happy to see us, and it was obvious that Harry had taken good care of her while we were away, but she seemed subdued and quieter than usual.

Since Ben was due to start the third grade in early February, I took him to Loneragans, the all purpose local store, to outfit him for his new school uniforms. That February was unusually hot with temperatures hovering around 38 C (100.4F) and I felt sorry for the school kids in the classrooms without air-conditioning.

But far worse was yet to drift up from drought ravaged Victoria. On 8 February, unusually high temperatures of 43.2C (109.8F) hit Melbourne in what was the hottest February on record. For a week, Victorians sweltered in the heat wave until a rapid temperature drop caused gale force winds to uproot trees and destroy homes. Next, a massive dust cloud, thousands of meters high, carrying 50 million tons of prime farming soil, swept into the city and obliterated all visibility. The eerie dust tidal wave could be seen from space, as it rolled eastward along the entire state of Victoria and into New South Wales.

A week later, on 16 February, wild fires erupted, first in South Australia and then eastward through Victoria. Freakish wind conditions fanned the fires and formed raging firestorms that generated enough heat to explode homes, melt steel and liquefy glass. The death toll of Ash Wednesday fires came to 75 people. Livestock losses included 340,000 sheep, 80,000 cattle and countless native animals. Over a million acres were consumed by the fire and thousands of dwellings destroyed. This was only the beginning of the worst drought on record that had just begun to affect our land in the NSW tablelands. The dark side of nature was about to be unleashed and it would bring challenges that no one had any prior experience in managing.

Ever the optimist, Bill decided to pasture-improve the paddocks with clover and sow winter oats for stock-feed, on the chance that it would rain. When he had finished, it stopped raining altogether. Each morning after Bill thought it had rained overnight, he raced out to check the rain gauge on the front lawn, but it had been just the wind on the corrugated iron roof. Sometimes only a few drops of rain teasingly fell from a passing cloud, and eventually we counted on the morning dew to keep the grass alive. We constantly searched the sky for a sign of rain and found nothing but the relentless sun in a cloudless sky. The oat and clover seeds we had sown lay un-sprouted, while the caustic un-dissolved super-phosphate granules scorched the foliage around it. On our morning rounds in the Rover, nothing escaped Bill's notice.

'Look at that phosphate burn', he would say. 'Look at that leaf curl'.

We drove past the water dams that Bill had excavated to drought-proof the property, only to see a puddle of muddy water encircled with fissured mud and imprinted with hoof imprints, not only from our own sheep and cattle, but from kangaroos coming down from the National Park searching for food and water. Even the dams that were fed by underground springs had falling water levels, an ominous sign that the underground water table was dropping and that this was building up to be a far-from-ordinary drought. Nature was starting to spin seriously out of control.

It began slowly, almost imperceptibly, after the 1983 summer grain harvests failed in January and the winter seed grains lay un-sprouted over the grain growing regions of southern and eastern Australia. A few dozen mice scampering around our silo and hayshed turned into a few hundred overnight, and then exploded into thousands, until millions of rats and mice had invaded our district. They moved in grey waves over the driveway, hay sheds and pastures, devouring everything in sight. We drove over them on the roads and stepped on them while walking. Our farm cats soon had their fill of mice. They gave up catching them and only chased them out of their immediate territory. The grey marauders invaded our home, ate our household appliances, and chewed the electrical wiring in the house. Each time our fuses blew, I knew that a rodent had just fried himself. They even gnawed on our plumbing, including the old lead kitchen pipes. Rodents were eating our house and contents as if it were nothing more solid than a loaf of bread. I was surprised to find that white goods had edible parts in them; the oven insulation was a particular favourite. Each morning I was greeted by the demolition wreaked in the kitchen overnight; fiberglass wool stuffing protruded from the oven, and particle-board cupboards were reduced to slivers.

One morning I put a piece of bread in the toaster, and after a few minutes the toast popped up, complete with a smoldering mouse. The sight of a charred rodent was bad enough, but the smell was too much first thing in the morning. I fled into the shower, hoping the problem would go away by the time I was dressed. It didn't, judging by Ben's screams when he went to get his breakfast. The entire house had a mousy smell that couldn't be shifted as long as the rodents were living with us. I tried trapping them, poisoning them, cornering and whacking them with the fireplace poker, but nothing eradicated the pests and no one, not even the Department of Agriculture, could tell us when, or even if the plague would end. I worried about the health hazard it posed to all of us, but especially to young Ben. I was repulsed by the problem and tired of having to kill mice. For the first time, I found conditions in the bush so hard to take that I wanted to escape the stress for a while.

To make matters worse, the rabbit numbers started building up. We tried culling them, and went out spotlight shooting with the .22, but it made no difference to the numbers.

One day the district rabbit inspector came onto our property from the Pastures Protection Board and announced that the district was now officially classed as drought stricken and that rabbits were decimating what feed there was left for stock and native animals. As a non native and invasive species, he said, the rabbits had to go. He told us that we were required to eradicate them with Myxoma virus from the glass vials that he had brought with him. This remedy made neither of us very happy. Bill didn't like being told what to do on his own property; he thought he was managing the drought as well as anyone in the district, and I wasn't happy about having a biohazard virus in the house. But since Colonial days, the Rabbit Nuisance Act of 1883 allowed inspectors the right to access private property and, to this day, the Act gave them power to compel landowners to comply with their orders.

The inspector told us the virus had to be stored in the fridge. That made me doubly nervous, since there was only one fridge on the property, and that was in my kitchen.

The next morning Bill handed the vials containing live Myxoma virus to the stockmen, whose job it now was to start an epidemic by catching a hapless rabbit, smearing it with a laboratory-cultured strain of Myxoma virus, and letting it go back to the rabbit warren to infect its community. At night the vials were cold stored, much too close for my liking, near the leg of lamb and the previous night's leftovers.

The resulting myxomatosis plague killed hundreds of rabbits and made just as many sick, but large numbers recovered from the virus with immunity.

Ben was heartbroken to see so many sick bunnies, and when he was home from school on weekends, he collected them in a shoe box and took them home to nurse them better. He had turned the space under his bed into a rabbit hospital.

Meanwhile, the rabbit inspector from the Pasture Protection Board (PPB) often called to check on our rabbit eradication efforts. One day, when he saw that the virus hadn't been effective in culling large numbers, he told us, 'Time to get out the 10-80, eh?'

10-80 is a lethal poison with no known antidote. Reluctantly, Bill and I drove the farm's Toyota Ute to the PP board in town to load up a ton of carrots laced with 10-80. We had provided the twenty-kilo bags of chopped carrots, while the inspector himself added the 10-80 and mixed it up in a rotary concrete mixer, right there on the premises.

As soon as we arrived home with the poison payload, the pressure was on to lay out the bait before the carrots went off. Bill and I helped Harry and the other stockmen move the sheep and cattle out to safe paddocks, before spreading the laced carrots around the pastures. We were careful to note the location of the bait so we could go back and meticulously pick up any uneaten portions after the kill. It was an impossible task to conduct a mass extermination of rabbits and ensure only the target species ate the poison bait. I worried about the native birdlife, but most of all I worried about Ben, and tried to keep him well away from the baited areas. I told him there were no second chances with 10-80, but it only made him more curious about the whole operation, and knowing Ben, he would have been feeling sorry for the rabbits.

The extermination was morbidly effective. Most of the rabbit population died after eating the bait, but the numbers of dead will never be known, whether they be native animals or rabbit pests.

The next day Bill and I took the Rover out to inspect the carnage in the paddocks. There was not much to see and it was strangely silent, but the stench of death wafted up from the silent rabbit warrens where countless thousands lay dead in underground warrens. Standing in the paddock after a mass kill made me feel empty and lonely and left a chill in my heart. There was no pleasure in grazing at a time like this, only a relentless battle with Nature.

Bill and I returned to the homestead for lunch that day, but I had lost my appetite. Nothing made sense to me anymore. Even the idea of having a baby seemed inadvisable in this hazardous environment. I had never seen Nature this far out of balance before and it rocked me at a fundamental level, as though I had lost trust in Nature itself to be able to provide for us humans. I served Bill a sandwich at his coffee table and sat down across from him.

'I need to get away from here', I announced, 'just a short break away somewhere'.

Bill looked at me curiously. I hadn't made such a direct request before.

'I can't go anywhere in this drought', he said.

'I'll go on my own then'. I answered. 'Ben can come with me. It's his two week school semester break next week'.

Bill looked at me disbelievingly. I gazed back at him steadily. Even I had found my outspokenness surprising, but I needed to breathe air that didn't smell of drought and decay.

On the first day of Ben's two week break I bundled him into my Honda station wagon and headed towards far north Queensland, to a resort owned by my friends on Magnetic Island, 8 km offshore from Townsville. Distances in Australia are vast and daunting. In hindsight, it was madness to drive a 4200 km return trip – the equivalent of driving from Barcelona to Istanbul or from Milan to Moscow – and squeeze such an excursion into just 14 days, but it was the best way to see the country close-up.

New South Wales was dry and brown almost all the way to the Queensland border when the landscape changed into lush green sugar cane fields that stretched all the way up the Queensland coast. The long, straight highways seemed endless, but Ben was taking it all in; from the sugar cane trains, to the banana plantations and finally, the tropical blue waters of the Barrier Reef. After four solid days of driving, we arrived in Townsville and took the car ferry to Magnetic Island.

My friends put us up in one of their thatched roof cabins. After a delicious dinner of grilled reef fish, we slept off the long drive in soft beds with no alarm clocks in sight. Ben spent the next four days making friends with the local kids and we all swam, basked in the sun and watched huge orange sunsets dropping below the tropical blue-green water. I spent the evenings in lively discussions with tourists from Europe and the US. The 5 days went by so quickly that Ben and I didn't want to leave. I wasn't looking forward to the long drive home or the scenes of drought and devastation on the way, but I was beginning to miss Bill, and starting to feel guilty about going off on a break without him.

The return trip seemed just as endless, until we reached the NSW drought zone again where I saw an alarming new development. The drought had become worse, even while we were gone, and now we saw bare soil paddocks with half-starved livestock searching for a blade of grass. Ben and I watched drovers on horseback droving sheep and cattle down the roadside in search of the feed that still grew there.

In the colonial days, the roadside was called the 'long paddock' when the highway verges doubled as stock routes that formed a nationwide network of corridors to transport animals to markets. Now the 'long paddock' was used to provide the last bit of feed to hungry animals, and we had to stop along the way to let mobs of sheep and herds of cattle cross the road to the other side.

On one of these stops I noticed my fuel gauge was low. Since we hadn't passed any buildings for a while, I realized that I must have taken a wrong turn at Sandy Hollow and mistakenly headed deep into the unfamiliar territory of the Bylong Valley. With no houses in sight, and no one to ask for directions, we were lost. And, in the days before mobile phones or GPS, being lost meant being lost. By now I was trying to decide if I should turn around and retrace my steps or carry on in hopes of coming out somewhere near a petrol station. The gauge now showed empty and I wasn't sure how much fuel, if any, was in reserve. I tried staying calm for Ben's sake, but it was already late afternoon, and spending the night in the car wasn't an option I wanted to take. I took a chance and drove on until I came to a cross road. There, I had to decide whether to turn left or right. Like a metaphor for life, either way would lead to a different destiny, and I was about to have a chance encounter with a long forgotten dream.

Finally, I turned right onto the paved road that promised to lead to some kind of building, and almost immediately, we found ourselves in a magical green and fertile valley. Ben and I were spellbound to see the wide and grassy plains sloping up the valley floor toward towering sandstone escarpments that shimmered yellow and pink in the afternoon light. The winding road led us deeper into the valley, along crystal clear running streams and grazing horses in lush Lucerne pastures. Despite the empty fuel gauge, I felt a blind urge to keep driving and wondered if we had gone through a time warp and emerged in some utopian universe. Later I discovered that the enchanted valley was once the home of the Kamilaroi Aboriginal tribe who named it Widden, which meant: 'stay here, go no further'. For eons the Indigenous tribes knew that this deep valley afforded permanent water, abundant food and protection from harsh weather and drought.

Ben and I were enthralled by the magical scene until he spotted a thicket of shade trees in the distance and excitedly pointed out a building in the middle of it. In a few minutes we parked in a circular driveway lined with white quartz pebbles, next to a magnificent ivy covered sandstone villa. I asked Ben to wait in the car while I knocked on the door to see if anyone would give us directions.

After several minutes, a man answered the door, looking surprised to see a visitor. After explaining our predicament, he gave me easy directions to the next petrol station and courteously invited me in to use the phone, in case I needed to let anyone know that I would be delayed.

With impeccable Old World manners, the gentleman led me to a plush timber paneled office where he showed me the phone and offered me a chair across from a polished mahogany desk. I openly admired the photos and oil paintings of the race horses that lined the walls, while he modestly acknowledged the compliment before leaving the room to give me privacy.

I dialed our home number while idly glancing around the office. As I waited for Bill to answer, I stole a closer look at the paper clipped to the board above my head. It was a stud service schedule, listing all the mares that were scheduled to be serviced, and the stallion who would service them. I felt guilty for being so curious, but when I saw the stallion's name, I was speechless. Meanwhile, Bill answered the phone and I felt myself being almost offhand as I let him know that we had taken a wrong turn and would be home later than expected. Still stunned, I hung up the phone. The stallion standing at the stud was a horse who'd had a slow start early in his career and was thought to be too small and unsound to race, but he showed himself a champion as a three year old and went on to sire some of the greatest race winners and broodmares in

Australian racing. His great great-grandsire was the American bred champion Man o' War, the greatest Thoroughbred racehorse of all time. Standing at this stud was none other than the legendary Bletchingly, and I had serendipitously wandered into one of the finest Thoroughbred studs in Australia – the Widden Stud, owned by the Thompson family, whose members have bred racehorses in the famous 'valley of champions' for a hundred years.

Half an hour later, with Mr. Thompson's help, we were on our way toward a petrol station and then homeward bound. One of my dreams had come true, when I stumbled upon the home of the legendary horse, whose story of victory, after being a late starter, had inspired me to never give up in my own life. I had also learnt that sacred areas existed in the world, and I was granted the privilege of seeing such an ancient haven, safe from droughts and floods.

We arrived home that night, in a good mood and just in time to save Bill from overdosing on cholesterol. His culinary skills were limited to greasy sausages and soggy chips and I could see that he was pleased to have me back. He was particularly proud of his housekeeping skills, since the mouse smell had subsided and the cooler weather had stopped the mouse plague almost overnight, as quickly as it started.

Even though science could not explain the mystery of the mice and rabbit plagues, it was Nature who was delivering a message, but few were listening.

Happy News

> You are the bows from which your children as living arrows
> Are sent forth.
> The archer sees the mark upon the path of the infinite,
> And He bends you with His might -
> That His arrows may go swift and far.
>
> *-Kahlil Gibran*

My good mood lasted as long as my holiday tan. After dropping Ben at the bus-stop, I came back to hear all of Bill's bad news, over breakfast. Since the drought had settled into NSW the animal death toll was broadcast every night on the news, but Bill had worked hard to avoid any stock fatalities on Landfall.

On my first trip out in the Rover since coming back, Bill showed me how he had trimmed our stock numbers to match the pasture carrying capacity. Even though it meant selling unfattened young stock through the sale yards and getting low prices, he bit the bullet for the sake of land conservation and it showed; our property was still greenish. Bill also filled up the silo with 200 tons of feed oats and worked out a way to irrigate the Lucerne paddock from an underground spring. This emergency supply of stock feed would take the pressure off the pastures and keep our breeding stock in good condition. Clearly, not every property owner could afford these expensive measures and Landfall became an oasis compared to the ruined properties all around us with hungry animals standing on bare soil. Some desperate graziers even had to euthanize their staving animals and bulldoze their livelihood into lime-pits. As I had seen for myself on the trip, I told Bill about the graziers who'd saddled up their stock horses, packed their swags and headed their sheep and cattle up and down the dusty long paddock for the dry grass that still grew there.

Since Bill had worked so hard to keep our stock in good condition, we were all the more shocked to discover a few dead lambs in the paddock during our rounds one morning. I went straight into vet mode and insisted on collecting one of the bodies for an autopsy. Bill obliged me because he hadn't seen animals die so strangely before with their necks arched back and faces distorted into grotesque grins.

I shoved one stiff lamb's body into a plastic bag and we drove it to the vet in town; the same one who had treated our dog a year earlier.

'How's your dog?' asked the vet, who remembered Polly's accidental poisoning.

'No good', answered Bill. 'She used to like running after the Land Rover for miles, but now she just lies around on her mat'.

'That can happen after organophosphate poisoning. Lucky your dog even survived', said the vet, as he started to look over the dead lamb.

'See how its neck is arched back?' asked the vet.

Bill and I nodded. We didn't think an animal could extend its neck that far back.

'Well, that's a textbook case of tetanus', said the vet. 'I won't even have to do a post mortem', he added.

'But we vaccinate against tetanus during the lamb marking', I protested.

'That may be so', he replied, 'but maybe this one got it before he could make antibodies'.

'What else can we do?' I asked exasperated.

'Nothing much, bar making it rain to keep down the dust...and keep on vaccinating. Believe me...', added the vet, 'you're one of the lucky ones, if you still have feed'.

It sounded as if the man was worn out from putting down drought affected animals and thought a few lamb losses paled in comparison. I now felt pressured to find a solution to this tetanus outbreak myself, but I hadn't dealt with an outbreak before and didn't know how it would unfold or how many more lambs it would claim by week's end. Science had taught me that disease outbreaks did not happen in 100% vaccinated populations. It blamed any outbreaks on unvaccinated individuals. Since I had believed this wholeheartedly, I could hardly believe that this outbreak had happened at all, when we vaccinated all of our animals on a regular basis. I still recall those heated arguments between factions who believed that improved public health measures were responsible for eradicating common infectious diseases, and others who claimed it was the advent of vaccinations. In the past, I had always agreed with the vaccination advocates, but in this case, what could I do when we had already vaccinated 100% of our animals for tetanus and four other common sheep diseases? It was now clearly obvious that outbreaks could happen in fully vaccinated populations and that I would have to find an alternative way of stopping it.

On the way home, Bill drove the Ute while I tried to reason through the whole problem. At lamb marking time the mob is mustered into the yards where the lambs are separated from the ewes. One by one, each lamb is then strapped into a wire cradle with its hooves pointing upwards. Before the animal knows it, two pairs of hands have issued it with a tail ring, a scrotum ring, 5 vaccinations, and an ear tag for identification. Its scrotum and tail area is then painted with sheep dip (OP pesticide) before its tail is cut off for good measure. (Graziers dock sheep's tails to prevent fly strike and they castrate male lambs to prevent inbreeding the flock.) After marking, the lamb returns to the ewe feeling justifiably sorry for itself and searches for the comfort of her udder. The next day the lambs have recovered from the procedure and 10 days later the rubber rings cause the scrotum and tail stump to drop off, while the vaccinations are intended to prevent tetanus, pulpy kidney and a host of other diseases. That meant that the problem must have come from the open wounds exposed to the dusty manure-ridden environment in the stockyards. I knew from my training that this was an ideal breeding ground for tetanus and if we could prevent the infection through open wounds we might be able to stop the outbreak.

By the time we arrived home, I ran my theory past Bill, who put up no objections when I asked him if I could talk with the farmhands about it. I thought I was onto something, and went to look for Harry.

'Why do we cut off lamb's tails when the rings make the tails fall off anyway?' I asked.

Harry stroked his chin while considering my question. 'So as we can count how many lambs we've done by the number of tails', he replied, no doubt wondering why I would ask.

'What if we leave the tails on from now on, let the rings do the job and find another way of counting how many lambs we've marked', I suggested, trying not to sound too forward.

Harry walked off to reflect on the problem. It was the first time I'd made a direct suggestion and I wasn't sure how it would fly in the strong chauvinistic tradition that still existed in the country regions. But it was worth a try, since the lambs had a better chance of surviving the dusty stockyards if they had no open wounds through which the tetanus to enter, especially since the rubber rings would remove the tails in 10 days without an open wound.

Meanwhile, the batch of lambs we had marked 10 days previously were now dropping dead by the dozens in a time frame that exactly corresponded to the incubation period of tetanus. Bill and I were heartbroken to see the ewes standing by their dead lambs in bewilderment.

To his credit, Harry agreed to try my suggestion during the next lamb marking. Before we started, I raised a few eyebrows when I wet the yards down with precious water from a nearby tank to reduce the dust. The tails were ringed, but not cut, and from that day on the tetanus outbreak ended as suddenly as it had started. Soon after, I arranged for all of us humans to have tetanus boosters as well.

Despite all the animal health agents we had at our disposal, it turned out that a simple common sense hygiene measure stopped the tetanus outbreak. This was a radical thought for someone like me, who had been deeply conditioned by western science and was almost evangelical about its proclaimed powers. But my beliefs were being challenged by what I was seeing and experiencing and it became increasingly obvious that no high technology could exert control over an out of balance environment or restore it to equilibrium. Nature had the final say, she would teach us what we needed to know, and she clearly was not happy. After being in Widden Valley, I also began to understand that the Indigenous culture knew far more about environmental balance than western science did. The Aboriginal tribes have known for tens of thousands of years, or more, which land would provide for all their needs, even in times of scarcity, and they knew how to live on the land harmoniously. Despite my growing interest in Aboriginal culture, information about it was hard to find, since it was not a culture of written language, but one that imparted its wisdom only to those who would develop the ability to listen and tune in to the 'dreaming'. I had hoped that one day I would be able to hear its timeless whispers and as it turned out, I didn't have too long to wait.

The first of June 1983 started with a vivid dream that gripped me tightly and made it difficult for me to awaken from it. I dreamt of a tribal elder who placed his finger on my forehead and 'downloaded' information of such stunning clarity and importance that I tried hard to remember it, but it all dissolved as I emerged from deepest sleep into my normal consciousness. That morning I woke into the peaceful world of 1983 where the most dramatic news of the entire month was the reelection of Margaret Thatcher and India's victory over the West Indies for the Cricket World Cup. But for us, it was no ordinary day. This particular morning I felt sick and not up to sitting at the breakfast table with Bill.

After taking Ben to the bus, I bundled up in Bill's sailing anorak and told him I was going for a walk, up to the ridge, for some fresh air. He looked up from his calculator and shook his head. I couldn't understand why I felt compelled to take a hike up there either, and didn't try to explain it.

On that crisp winter's day, I so enjoyed standing on the sun drenched ridge that I took off my shoes to be closer to the earth. Gazing out over our homestead in the valley below brought up feelings of deep gratitude for everything in my life, as though my deepest wish had been granted. As I breathed in the fresh air and felt the warm winter sun on my face, I felt a sense of oneness between myself and nature, and the all powerful force of creation. In that transcendental moment I knew, beyond a doubt, that our lives had changed, and with an explosion of exuberant energy, I raced home. Bill was speechless when I burst in the door, grabbed my bag, and left again to jump into the car and head into town. An hour later I came back, only to disappear again for a few minutes, before joining Bill at the table. He hadn't moved from his place where he was calculating stock numbers, except to fix himself another cup of coffee. But now he looked at me as if I were a mad woman.

'What's up with you?' asked Bill.

I couldn't hide my Cheshire cat smile any longer. 'Looks like you're not going to the knackery after all'.

Bill looked me in eyes to gauge my seriousness. 'You're kidding me. Are you...'

When he saw that I was serious, he got up to envelop me in a long bear hug. 'Thank goodness I won't have to have any more of those tests!'

'No, but you'll have to change nappies instead', I retorted.

Bill grimaced at the thought of a dirty nappy, and then smiled. 'I hope the baby has your looks and my brains', he answered, hoping to get a rise out of me.

When I picked up Ben that afternoon, even he sensed something was going on and asked why I was so happy. As I'd predicted, Ben was thrilled about the good news. He had wanted a sibling as a companion and didn't mind that the baby was going to be nine years younger.

'When do you think the baby will be old enough to play?' he asked. Something told me Ben wouldn't miss being an only child.

The good news would also bring Ben some unexpected fringe benefits. Every 4-6 weeks we would need to commute to Sydney for a long weekend so I could visit the obstetrician who would monitor my pregnancy and deliver the baby. Meanwhile, Ben would get to hang out with his friends at Bayview, as a welcome break from the bush.

We did the shearing when I was four months pregnant. I was determined to keep helping Bill for as long as possible, and had no expectations of being pampered, since Bill wasn't exactly the pampering kind, despite his happiness about being a father again.

It seemed to me that August, being the middle of winter, is a peculiar time to rid sheep of their woolen coat, but Bill assured me it was the done thing, since sheep liked to have their coat on during the summer when it afforded insulation. I wondered if anyone had ever asked the sheep about that.

If I thought the sheep had it rough, then the shearers were not far behind. Each morning at 7.30 sharp they turned up to the shearing shed in their blue singlet and baggy jeans, held up by a belt saturated with wool grease. Shearers typically are thin as greyhounds, with a bent back, flat backside and muscular forearms. They are custom built for catching sheep in the pen, dragging it onto the shearing board and clipping it with electric shears - belly wool separately - until the fleece comes off in a whole piece. Over the nine grueling hours of bending over, the men are in fierce competition with each other for the highest tally of shorn sheep, but revert to best buddies before heading home at night. Shearers know very well that most people couldn't do a clean job of shearing even one sheep, let alone the 200 sheep a day that a 'gun shearer' can shear.

My job in the shearing shed was to help Harry, Warren and a few contract rouse-abouts to pick up the fleeces, spread them on the revolving wool-classing table and 'knock the dags off em'. This meant that the manure and urine-stained wool around the edge of the fleece had to be discarded so the wool buyers would not penalize the whole fleece for organic matter, and the wool could be uniformly classed according to its fineness and crimp. The belly wool was bagged separately, since it was price discounted for its dirt and vegetable content.

Bill was wearing his wool classer's hat and after evaluating each fleece he delegated it to various piles, marked A or AA or AAA, with triple A being the finest wool. Most of our flock consisted of Border Leicester–Merino cross, which produced excellent fat lambs, but large amounts of medium quality wool that averaged 25 microns in thickness. Our fine wool clip came from our 4000 pure bred Merino sheep which consistently produced fine quality AAA wool that ranged between 19 - 21 microns. Since our wool clip made up 30 per cent of our yearly income, Bill was always laser focused on wool classing and most knew better than to disturb him on the job.

At 12 o'clock sharp everyone stopped for an hour lunch break, except me, since I was expected to cook and have lunch on the table for Bill. An hour later, just when I was getting comfortable on the sofa, Bill and I returned to the shed for the afternoon's work. Between three o'clock smoko and knocking off time, the shed turned rough as the shearers felt their backs straining. 'Come here, you bloody bastard!' they would grumble, before locking another sheep into a half-nelson.

Harry picked up on the edgy atmosphere and gave the rouse-abouts a hammering whenever they slowed down, while Bill cursed the wool press when the wool bales were underweight.

By 4.30 I brought Ben home from the bus, and dropped him at the shed to lend a hand. He was wise to jump into the mountains of wool and watch the action from his safe, fleecy hiding place until 5.30 pm when the light faded and everyone left the shed in a mass exodus.

'Oim goin for a cold tinny and a hot woman - in that order!' The shearers shouted out their ritual joke, in unison, before lighting their cigarettes and driving home in their bent up Utes.

The work went on for two or three weeks, until every shorn sheep was returned to the paddock, and every last bale of wool was loaded onto the semi trailer bound for the wool broker's warehouse. After that, the wait was on, until Bill got the wool cheque.

That year Bill was worried about the state of the wool industry. Since February, 1983 the wool market had suffered such a severe downturn that 4.75 million bales were stockpiled in warehouses without buyers, after the Soviet Union, a major importer of Australian wool, began its economic collapse (this led to its reemergence as the Russian federation in 1991).

Since then, Australian wool sales have declined almost yearly. Australian graziers and wool producers have repeatedly asked why exports are declining in an industry that Australia had once globally dominated. Government and academics maintain that synthetic fibers are reducing the global demand for wool, but this does not explain the fact that USA and China are expanding their wool production, broadening their markets and finding buyers for their wool. Australian wool growers frequently complain that the biggest wool industry the world has ever seen has been sold down the river by repressive government policies, harmful political and social agendas, pressure groups, lack of representation and overregulation.

Six weeks later Bill got the wool cheque. As he had feared, it was half the amount it usually was. On top of the drought, the wool market collapse couldn't have come at a worse time.

But despite all the hardships of the drought, this year had also been good to us, with the prospect that we would be expanding our family. During our six weekly trips to Sydney, my doctor announced that my pregnancy was progressing very well and all was normal. Bill was also getting used to the idea, and we had fun thinking up names, while Ben waited patiently for his playmate.

Droughts and Yacht Races

Our Andy's gone to battle now
'Gainst Drought, the red marauder;
Our Andy's gone with cattle now
Across the Queensland border.

- Henry Lawson

Before dawn on 27 September, Bill snuck out of bed, slipped on his woolen dressing gown and installed himself in front of the TV in the winter room. I came out to join him, just in time to watch two sleek twelve meter yachts fight out the last half mile to the finish line at Newport Harbor, New York. Both our eyes popped out when we saw that Australia II, skippered by John Bertram, was seriously challenging the American yacht, Liberty, skippered by Dennis Conner.

Since the America's Cup was first held, 132 years ago, the yacht race had never been won by a foreign contender, and the thought of seeing the prized trophy leave the New York Yacht Club was enough to make American yachties tremble in their Sperry Topsiders. But there it was, for the world to see; Australia II was streaking toward the finish line, more than seven lengths in front of Liberty, and you could almost see Dennis Conner weeping at the helm.

By now, Bill had sprung up from the couch. 'Come on you bastard, Bob Miller', he shouted at the TV. I wondered who Bob Miller was, but now was clearly not the time to ask.

The noise had woken Ben up, who emerged from his bedroom rubbing his eyes, just as Australia II made history gliding past the finish line. Her easy victory was thanks, in part, to her secret weapon below the waterline, the winged keel. By then we were all glued to the TV as we watched the winners arrive at the press conference, looking slightly dazed, as if their victory hadn't yet sunk in. The euphoric mood was infectious and compelling, and I was barely able to tear myself away long enough to make Ben's breakfast and drive him to the bus-stop.

When I returned home, Bill was still watching the media aftermath, including footage of Prime Minister Bob Hawke's famous statement: 'Any boss who sacks anyone for not turning up today is a bum'.

Interviews with skipper John Bertram, owner Alan Bond and Australia II designer Ben Lexcen followed, this time they looked more relaxed and Lexcen looked amused when Prime Minister Hawke told an American journalist, by satellite link, that the America's cup would be in Australia for a 'very long time'. Bill laughed at that brazen statement, since he had the same competitive streak in him. But there was one thing that I was still confused about.

'Why did you talk about Bob Miller earlier?' I asked Bill. 'There's no Bob Miller in the group. Who is he anyway?'

Bill gave his amused high pitched cackle.

'Bob Miller changed his name to Ben Lexcen', he explained. 'And I know him well, along with most of the others we saw on TV this morning'.

For the next half hour Bill told me the fascinating story of his collaboration with Ben Lexcen (Bob Miller) while designing the third in a series of Bill's own ocean racing yachts, a 40 ft sloop launched in the 1960s, especially to challenge for an international yacht race in England. Bill had designed and built yachts before, and successfully raced his previous boats in several Sydney Hobart races, but to take on the tough international challengers, he needed a design edge that produced a lightweight and maneuverable racer that still adhered to the strict RORC rules regarding hull design. After Bill worked out the design challenges on the drawing board, he appointed one of the best master boat builders available to produce a light weight racing boat; the first of its kind in Australia, made by fusing two or more thin layers of timber that run in different directions, to give a strong load bearing outer hull that does away with the need for too many heavy load bearing internal structures. The result was a sleek, lightweight and strong yacht with a revolutionary design and construction. Bill would skipper her and hand-select a crew to challenge for the most coveted yacht race in the world.

After loading the racing yacht onto a cargo ship, he worried about her for a solid two months until she arrived on the other side of the world, just in time for the start of the ocean race. He needn't have worried though, after her long sea voyage, she didn't miss a beat and went on to dominate the trials, winning 9 firsts, 1 second and 2 thirds in the first 14 races. Naturally, the race committee had to select her as a contender for the race, and she did not disappoint. With Bill at the helm, and supported by a seasoned and capable crew, he ended up being the highest point getter and won the race by an impressive margin.

'What happened then?' I asked Bill.

'Then we went to dinner with Prince Phillip', he told me matter-of-factly.

'You're kidding me', I said breathlessly.

'The invitation from Buckingham Palace is somewhere at Bayview. See for yourself', said Bill.

'I sure will!' I exclaimed. 'Did you go back another year?'

Bill chuckled. 'I tried for it again, but came second. After that, I sailed in about the next 10 Sydney Hobart races'.

Bill had always enjoyed telling sailing tales about his yachting friends, but now I was beginning to better understand his passion for boats and sailing, and the real meaning of the precious contents of his trophy cabinet at Bayview. Now I knew why he felt so disheartened about not having a boat, especially on Boxing Day when he watched his sailing mates cross the starting line without him.

'Next time we're at Bayview, I'll dust off your royal invitation and hang it on the wall'.

Bill roared laughing and grabbed his black leather bush hat.

'Where are you going?' I asked.

'We haven't been out in the paddocks since the start of shearing. God only knows what they look like without the spring rains and with the summer heat coming on'.

To me, it seemed surreal to be wrenched away from the euphoria of winning yacht races to go driving around dry, dusty paddocks, and it must have been even harder on Bill.

I was now nearly six months pregnant, and although I told Bill I would no longer be doing any heavy work, I still enjoyed coming with him in the Land Rover.

Over the next two days, Bill inspected every paddock on the place and came back to the Land Rover each time, looking more despondent than before. The previous month of drought had taken its toll. When Bill noticed bare earth patches staring to form, he knew it was time for drastic action; around a thousand head of stock had to come off the paddocks to ease the pressure and preserve the condition of the remaining stock.

The following morning Bill instructed Harry and Warren to sell 700 lambs and feed out a thin trail of oats to the remaining stock every other day, while he returned to his desk to find agistment for at least half of our cattle. Finding someone with green pastures, who was willing to take our stock until it rained, wasn't easy in a drought that spanned most of Australia, since those who were so lucky, didn't advertise it. And provided Bill was fortunate enough to find such a place, he knew that it wasn't going to be cheap to agist half our herd of 600 Hereford cattle. But at least with half the stock off the property, the remaining 300 cattle and 8000 sheep would be able to stay at Landfall and the pressure would be off the pastures for a while longer.

When Bill still had no luck after the fiftieth phone call, he announced that he had no other option but to go 'walkabout' in hopes of finding a green property and an owner willing to agist our stock until the drought broke.

I was six months pregnant when I kissed Bill goodbye and watched him disappear down the road in his smoky old diesel Mercedes. He first headed toward the Victorian Goulburn River floodplain where some of the largest irrigation systems in Australia are installed. Bill thought it was a good bet to find a patch of green there. Following that, he intended to check out the Queensland/NSW border after I had told him about the green country I'd seen up there on my way back from Townsville. There was no telling how long he would be gone.

While Bill was gone, Harry and Warren mustered in the herd of horses from the back paddock. They all had to be wormed and vaccinated while the Anglo-Arab yearlings needed to be handled, broken to halter and branded. Since I was later in my pregnancy, I stayed away from the dangerous work inside the yards, but was still kept busy allocating the correct branding irons to the right horses; each needed a station mark branded on their shoulder that included a number designating the year and their rank in the herd. I also arranged for the farrier to trim the horse's hooves and collected the vaccinations and drenches from the vet supply store in town. Among the group were several colts that needed gelding since the stallion, Shimaal, wouldn't tolerate other males in the herd when they were past a certain age. In those days, station hands did castrations in the stockyards, and Warren took care of that, with a few contract farmhands.

Meanwhile, the purebred mares in the front paddock needed to be serviced and I had to arrange for their transport to Victoria to visit a purebred stallion there. Although we were up for a stiff service fee, at least a few more mouths were off the property for a while.

While Ben and I were having dinner one evening, Bill phoned with some encouraging news. On his marathon two week trip from NSW to Victoria, and all the way back up to the Queensland border, he had found a property outside Sheppar-ton Victoria, with plenty of feed and an owner who agreed to agist 150 of our cattle.

After that, Bill headed up the NSW north coast and found another place at Grafton for the other 150 cattle. He asked me to let the men know; and by the way, would I please phone around for double-decker cattle transports, he added. Bill was accustomed to issuing instructions to anyone, even his wife, when a job needed doing.

The next morning, after I told Harry and Warren the good news, they headed straight out in their Ute to start a full cattle muster. With the help of several sturdy Australian Cattle Dogs (Blue Healers) they would spend the next few days separating out 300 cattle from the herds in various paddocks. Cattle Dogs are known for their intelligence, courage and loyalty and the only breed of dog with the guts and stamina to dart into a thundering herd of cattle and come out again alive. The men and dogs would soon bring the 300 Herefords into the 20 acre stockyard holding paddock where they would be rested, fed and ready for loading when the cattle trucks arrived.

That didn't solve my problem, though, since booking cattle trucks was no easy matter. Every grazier, who could afford it, loaded his herds into the countless cattle trucks that rolled day and night around Australia looking for green feed, and there weren't many spare trucks available. But I knew how important it was to Bill and I gave it my best shot. Our herd of prime Hereford cows and top bulls weren't a commodity that he could sell and buy back when times were good. When the drought broke, good stock would be impossible to get, or prohibitively priced. It had taken Bill 20 years to build up a quality herd, and he didn't want to lose his life's work, if he could help it.

Two days later Bill turned up. I was almost seven months pregnant and hugely relieved to see him again. I'd not only struck out with the cattle trucks, but I'd worried about what I would have done if I had injured myself, or if the baby had arrived early, and dreaded the thought of having to phone a neighbour, or drive myself to the hospital, since neighbors weren't close by. It surprised me that, even with high technology, my predicament was little different from the pioneering women who'd lived in the region a hundred years ago when their men had to find greener pastures for their cattle in a drought.

As always, Bill had a lucky streak and a few days later, five double-decker cattle transports rumbled into our stockyards, down the hill from the homestead.

Later, Bill and I watched the men loading our 300 head of cattle, up the stock loading ramp, and onto the five livestock trucks. The transports took 31 head per deck, leaving an adequate amount space between animals during the grueling trip.

After a days' loading, the heavy 18 wheelers rolled away into a thick cloud of dust with our precious breeding stock aboard. I hoped that the animals would all get to Victoria and to the NSW north coast alive. Since the drought, I felt sorry to see the stock transports on the road, shifting hundreds of thousands of cattle to greener pastures. Stock wilted in the heat of the day; they had to be let out and watered, so the truckies drove like the devil each night until they got to their destination. You didn't want to get caught in the slipstream behind a cattle transport or road train. They moved so fast you couldn't pass them, and they left a trail of stench for miles. The road to somewhere green was long and miserable, but there was no other choice in a serious drought like this.

Bill turned to me when the dust had settled. 'I wonder what Allan Bond is doing right now'.

'Probably not this!' I replied.

His thoughts were never far from owning, designing or sailing yachts, and just then, I couldn't blame him.

Innocence Lost

It is a poison that disintegrates
Friendships and breaks up pleasant relations.
It is a thorn that irritates and hurts;
It is a sword that kills.

- The Buddha

Just when Bill and I thought we could soon take a well earned rest in Bayview and look forward to our new arrival, Harry knocked on the door one afternoon to hand in his notice. Officially, it was because of his bad back, but the resignation also included his son Warren, whose back was fine.

From the winter room I heard the conversation, and Bill tried every tack to persuade Harry to stay on with Warren. He even offered them more money, but evidently it wasn't enough. A few weeks later we would find out why. For us, their resignation came at the worst time, just before the main lamb marking season. We had planned to leave for Bayview soon and now we had no farmhands for the lamb marking and no manager to oversee the property over the holidays, and during our baby's arrival. We were left completely in the lurch.

After Harry left for home that afternoon, I came out of the adjacent room.

'I think Harry's still smarting over not getting a Christmas Bonus', speculated Bill. He had given both men a bonus of $600 in the previous years, but over the last two drought years, the property had cost much more to run than the income it had produced, since we'd had less stock, lower carrying capacity and higher costs.

'Employees probably don't understand that', said Bill, 'They think we can afford anything'.

Bill was right about that, but he had seriously underestimated the work place market. He'd always paid the stockmen above award in wages, but some of the other large landowners in the district were digging deep and offering big money to good stockmen who could skillfully manage stock in drought conditions. A week later, Harry and Warren left, without any ill will. We never heard more about Harry's back and he never made a compo claim. But we did hear that Harry was managing the Grey's 10 000 acres and was paid a top wicket.

Despite a solid week of phone calls, Bill found no-one to fill the two full time positions. All the manpower in the district had dried up except a couple of inexperienced part time farm hands who Bill snatched up to help us with the sheep muster and lamb marking. That meant all hands on deck and Bill and I would have to pitch in with the heavy work. I had a bad feeling about it and wanted to stay away from the farm work at this late stage of my pregnancy, but there seemed no other option at the time.

I was seven and a half months pregnant when we began the sheep muster and lamb marking. That November turned out to be the hottest and driest on record. Maximum daily temperatures nudged 40°C (104 F) every day for a week.

After Polly was poisoned the year before, I made sure she was nowhere around. I needn't have worried; she had spent the past year listless and sprawled on the doormat.

Before starting the lamb marking, I read the instructions on the sheep dip. The label gave the mixing directions and a warning that read: *'Use as directed, wash skin after using, irrigate the eyes if necessary, and if ingested, seek medical advice immediately'*. I assumed the chemical was safe when used as directed.

While Bill mustered the sheep in with one farm hand, the other contractor and I prepared the marking area. We mixed the chemical in a container and added the required amount of water. The sheep dip contained diazinon, an organo-phosphate pesticide compound that included toluene, a vaporous solvent with a strong chemical smell that turned milky when mixed with water. Before the internet, there was no way of knowing anything about the chemical's effects on humans. Only hospitals had access to that information. I could only follow the directions and assume it was safe.

After the men mustered each new batch of lambs, we all worked together lifting the lambs into a wire cradle where we tail ringed, castrated (with rings) and earmarked them. I injected them with vaccinations and painted the sheep dip onto the lambs' groin and tail area with a paintbrush. Some of the sheep dip inevitably landed on my skin and clothes. There was no instruction on the tin about protective clothing.

Meanwhile the ewes were separated into the large race where one of the farmhands drenched them. After finishing with each batch lambs and ewes, we reunited them in the yards before Bill and the other farmhand drove them back to their paddock and mustered a new mob into the stockyards. This went on from morning to evening; the intense heat, the flies, the backbreaking work, the fumes and the sheep-dip all blurred into one long, hard day.

That afternoon, after I'd collected Ben from the bus, I had a headache and only enough energy to wash off the dust and dirt in the shower and go to bed. I put it down to hard work, the sweltering heat and pregnancy. Thankfully, Bill took over in the kitchen and managed to make dinner for Ben and himself.

The next morning I felt better, and although I could think of nothing worse than fronting up in the stockyards, there was no one else to help and Bill looked grateful when I'd turned up in my Honda wagon. It was another scorcher, and another early night feeling sick.

By lunchtime the next day, I felt too unsteady and shaky to continue work and I was also was worried about the baby; he seemed far more active than usual and was making thrashing movements inside me. A thumping headache was also starting to spread rapidly from my neck to both eyes and sideways up each temple. I told Bill that I wasn't feeling well and was going home to lie down. He nodded and went back to work.

On my way up the hill, I could barely drive the car; my eyesight was blurry, my hands were unsteady and my muscles wouldn't stop twitching. I braced myself to get out of the car and walked unsteadily into the homestead. Polly looked up from her place on the mat and for a second I was reminded of her accidental poisoning when she was twitching and walking unsteadily before her collapse into unconsciousness.

After that, I must have lost track of time, along with the ability to reason, but I do recall a strange mosaic of events as the poison began disrupting my brain and neuromuscular function. The splitting headache was the most excruciating head pain I had ever felt, and I remember thinking, oddly, that I should wash my hair to relieve the pain. In the shower, my blurry eyesight narrowed to tunnel vision that made it difficult to see anything other than what was directly in front of me.

When a wave of nausea engulfed me, and a flood of saliva filled my mouth, I was too disorientated and un-coordinated to find the toilet bowl, and I vomited in the shower.

Later, I wandered around house in a haze trying to find my clothes. I remember the baby's twitching and jerking movements and the trembling of my own arm and leg muscles. Every system in my body seemed to malfunction at once. I even felt my own racing heartbeat in my head and chest. When a few uncomfortable contractions started, I worried about going into labour, but was unable to get in the car and drive down to the yards to call for help. The poison prevented brain cells from communicating with each other and I simply could not think coherent thoughts or command my muscles to take orders from my brain. I suspected that I had been poisoned and wondered; if that's what the chemical was doing to my brain, what on earth was it doing to my unborn child's nervous system?

As the neurotoxin was scrambling my brain, and shutting down my neo-cortex (responsible for memory and reasoning ability), I recall switching into a different kind of awareness that evidently didn't require my brain to function coherently. I had a clear intuition that something was destroying my body at a cellular level and that I had exceeded some kind of physical threshold and was now at the frontier between life and death. In the absence of normal brain function, some form of primal consciousness told me that my unborn child and I were in deep trouble and that we might not survive.

My instinct was correct, since up to 30% of organophosphate poisoning victims die in the first three days after exposure. (Wadia 2003)

I faintly recall that Bill had appeared at some stage and I asked him to take me to the hospital. By then, speaking became difficult, since my mouth constantly filled up with saliva and I couldn't swallow it fast enough. Bill seemed stunned for a while, until he thought of helping me into the car and giving me an empty ice cream container to vomit into, which I did numerous times during the trip to the hospital.

Unfortunately for me, the state government had slashed the hospital budgets right across the state, yet again. This meant that our local country hospital had closed half its beds and was staffed only by a few nurses and a local GP, with no emergency room specialist in sight.

When we arrived, I was still incoherent and Bill was in shock himself, but not for the reason I'd assumed. Only later did I learn that he had rushed his late wife to the same hospital only four years earlier, where she had died of a heart attack.

The doctor on duty that night saw that I was in late pregnancy, vomiting uncontrollably and in a daze. If he had taken a more thorough medical history, checked my pupils, and ordered a blood test, he might have discovered that I had organophosphate poisoning, and hopefully he would have given me the antidote, atropine. This would have decreased my risk of dying, either immediately, or in a delayed reaction that sometimes occurs within 48 hours of poisoning, and lowered my unborn baby's risk of being stillborn or his chances of suffering permanent damage from the poisoning.

As it happened, the doctor assumed that I was merely dehydrated from vomiting and immediately inserted an IV, before giving me an injection of an anti-vomiting medication and a pain killer. He admitted me to the hospital ward, but without the antidote, I remained in severe pain from the headache and the baby continued thrashing around, which prevented any sleep that night.

That evening, I remained in a haze and don't recall seeing Bill or Ben. Meanwhile, no foetal monitoring was done on my baby; I was not transferred to a larger hospital with acute care facilities, and not referred to a specialist or advised to have my obstetrician check on the baby's welfare.

The next two days were a blur of exhaustion and pain. My muscles still weren't working smoothly; I remained unco-ordinated and unable to think clearly.

After the second day, Bill collected me from the hospital and offered to take me to lunch in town, but I just wanted to go home to bed. In reality, I was still as ill as the night I'd arrived at the hospital, since the neuro-toxic poisoning had been left completely untreated without an antidote, and would keep damaging the liver and nervous system of my unborn child and me (Koviacic 2003). Before the internet, I had no chance of finding out for myself what damage the poisoning had done and what treatment we would need to avoid complications. The best I could do was to see my GP and obstetrician as soon as we arrived back in Sydney.

I didn't know it yet, but the poison had changed my body permanently, and without an antidote, it would change the destiny of our unborn child.

I spent the following week in bed. Working out of doors was out of the question, but I even found it difficult to do light housework such as cooking and vacuuming. I asked Bill if he would agree to hire someone to come in once a week to help with the housework, but he refused. I asked him why he had provided household help to his late wife Sally, but not to me when I needed some help.

'She was a lot older than you', answered Bill.

'So, is that all I am to you, a worker?' I shot back.

Bill looked exasperated. 'You don't know what I'm up against here. We can't even leave this place until I find a manager to look after the property'.

It was the first time we had seriously argued.

'And you haven't heard a word I've said'. I replied. 'I'm your wife, this is your baby, and we need your support right now'.

Bill didn't want to acknowledge how sick I was, or admit that the baby might have been affected by the chemicals. He was completely overwhelmed by the challenges on the property and his old-school ideas kicked in: in his mind, he was a good husband and father if he kept the roof over our heads.

One afternoon, Bill announced that he was going to interview a stockman for the manager's position. The man had heard, around the local pubs that we needed someone with experience, and phoned Bill to ask about the position. Bill jumped at the chance and invited him to the homestead for an interview.

Allen arrived and greeted Bill with a confident swagger. He cut a lean figure in a pair of moleskins. From under his Akubra, a pair of steel blue eyes peered out of a red and rugged face. Bill hired Allen without references, since hired help was hard to find, and he thought he couldn't afford to be too fussy. When he left, Bill asked me what I'd thought of him.

'There's something strange about him', I said. 'I can't put my finger on it'.

Bill frowned as though I had just rained on his parade. He was relieved that he had, at least, found someone to take the reins for a while, and quite honestly, I couldn't wait to get to Bayview for a rest.

The following week Allen moved his wife and four kids into the workman's cottage, which sat half way up a hill, diagonally across from our homestead. His position as farm manager entitled Allen to rent free accommodation, as was the custom for managers in the bush, and Bill thought that sweetener would be Allen's incentive to do a good job.

Bill's remaining staff problem was solved when Darren, a young lad on an adjoining property, enquired about a job as a farmhand. Bill was thrilled that things were on track again and got busy showing Allen and Darren the ropes.

It wasn't long before Allen's wife, Barbara, came down to the homestead to introduce herself. I was still unwell and not in the mood for company, but she had brought me a large bunch of blooming yellow wattle that she had cut from the trees lining our road, and it cheered me up. I found a vase and we chatted over a cup of tea as two mothers. Barbara looked like a hard-working country woman under stress. She'd tried to freshen herself up with a little too much makeup but it didn't obscure the worry lines around her face. She already had a brood of four children but she looked at my pregnant waistline enviously. 'You're making me clucky again', she said, before a hint of sadness came over her face. There was a story there that I thought was best left for now.

'I'm so glad there's another woman on the place', I told her.

'Let's car pool the kids', she suggested. 'I'll see the next week out, till the end of this school term, and you can take over when the new school year starts'.

'What a great idea'.

The plan worked well. I could get some extra rest while Barbara drove Ben to the bus-stop, along with her kids, and dropped him back home at 5 pm.

At week's end, Ben came home with his end of year school photos again. That completed his second year in the bush and he was eager to return to Bayview and catch up with Peter and Jason.

In mid December, Bill felt he could entrust Allen and Darren to run the property and we loaded up both our cars for the trip to the city. Into Bill's car, we squeezed a month's laundry and Polly. The laundry came with us because I hadn't been able to wash in the bush since the mice had eaten out our washing machine. On top of that, the drought limited our tank water supply and I'd been too sick to take the laundry into the town Laundromat for the past month. Polly came with us, because Bill didn't want to leave her alone, since he realized that she hadn't been well after the poisoning.

I had my hands full getting Ben, our luggage and my eight month pregnant self into my Honda station wagon.

For the next four hours, we drove over the Great Dividing Range in convoy, looking like an eccentric caravan of nomads, in search of an oasis.

Destiny's Child

Your children are not your children.

They are the sons and daughters of Life's longing for itself.

- Kahlil Gibran

The green leafy streets of Sydney's affluent North Shore were an oasis compared to the drought baked bush, and our Bayview home was the sanctuary I had longed for.

I had always been strong and fit, but this time I struggled up the steep front stairs and had to leave the washing basket and suitcases in my car for Bill. I still felt so ill that I wanted to throw open the French doors and sit on the balcony with my feet up forever. But that indulgence would have to wait. Though I'd helped Bill with the farm work, he had not exactly embraced the idea of helping me out with the housework in return, so each time we commuted from one house to the other, an endless array of household settling in tasks were assigned to me.

Meanwhile, Ben wasted no time and took off down the road to look for Peter and Jason, and Bill installed himself at the dining room table with his calculator, working out the year's income and losses.

I opened the doors and windows to get rid of the mould that always settled in the house while it was unoccupied. This time, the mustiness gave me sneezing fits, an early warning that I was more sensitive than usual. It was time to make those doctor's appointments.

The next day, I relayed the whole story to my GP. The doctor looked blank when I asked him about the possible effects of the poisoning. He took my blood pressure and announced that all was well and ordered no further tests.

Later that week, my obstetrician announced that he could hear the baby's heartbeat. I was quite relieved, of course, but I knew the baby was alive, since he had continued to be hyperactive from the time of the sheep dip incident. Instead, I was worried about the reason for his sudden hyperactivity. The doctor seemed to think the baby's activity was a good sign and nothing to worry about. However, he expressed surprise that I had gained nearly five kilos in weight since last time he'd seen me, especially since I had not been overweight before. I explained that this had only happened in the past three weeks. At the time, I didn't realise that laying down fat was the body's way of getting toxic agents out of the way by suspending them in fatty tissue, nor did my obstetrician, who put it down to advanced pregnancy. The fatigue and lack of focus and concentration, he said, were due to my late stage of pregnancy. The vomiting, fortunately, had stopped. He thought perhaps I'd suffered from a bout of pregnancy vomiting in the third trimester.

Both my doctors were excellent and very capable when it came to the usual medical issues; in particular my obstetrician, who was also an exceptionally caring individual. But it would have helped me and my unborn child if all doctors had been more aware of the chemical risks to pregnant women. After all, chemical exposure is very common, since organo-phosphates and similar toxic chemicals are prolifically used in agriculture, industry, business and homes, and most people are exposed to these chemicals at home, in schools, at work, in household products and as pesticide residues in foods. Some people encounter acute toxic chemical exposures during pesticide fumigations, insect surface spray applications and agricultural use, and when they do, they expect their doctor to be able to treat them, since harmful effects of chemicals are thoroughly documented in the medical literature, and available to any medical professional.

Researchers have long ago established that organophosphate products can pose a 'significant potential health risk' to humans and that acute pesticide poisoning symptoms in pregnant women may mimic complications of pregnancy and they may be misdiagnosed and left untreated (Tafuri 1987). This was precisely what happened in my case, and the damage from this serious oversight is confirmed by researchers such has Solomon, who notes: *'Poisoning during pregnancy may result in serious adverse effects for both mother and the fetus or neonate.*

Prompt diagnosis and treatment are necessary to avoid adverse outcomes'. (Solomon 2007)

To prevent serious damage to me and my unborn child, special testing and appropriate treatments were urgently required. Even simple anti-oxidants, such as N-Acetyl-Cysteine, glutathione and vitamins E and C, would have protected us against the neurological damage and many of the DNA damaging effects that are known to be caused by this chemical (Shahin 2007).

Unfortunately for us, management of chemical injury is not emphasized in doctors' medical education and it is up to individual health care practitioners to refer to the Merck Manual or Poison Information when confronted by a patient with a history of poisoning or chemical exposure. This rarely happened in cases such as ours, and it took me another 11 years to find out why.

Meanwhile, since I knew that something serious had happened to me; as a patient, I felt entirely isolated and invalidated when no one acknowledged or treated the problem. Before the internet, I solely relied on medical professionals for help; but there would be no help from that quarter any time soon.

After the fruitless doctors' visits, my only option was to try and put it all behind me and prepare myself for the birth of my child. Compared to the past two years in the bush, life in Bayview was a breeze. My sketchy memory and brain fog now worked in my favour; I'd lost interest in most entertainments, including TV and women's magazines. Nothing tormented me in the final weeks before giving birth, and I didn't much care what happened in the rest of the world. Pesticide was far from my mind, though it was silently weaving its way into our lives.

Each day I sat out on the patio and gazed at the shimmering bay. Like a quiet meditation, the glistening water carried me into an inner sea of tranquility. On balmy summer nights I counted the flashes of the faraway Barrenjoey lighthouse that guided the ships at sea. Its steady beacon reassured me that at least some things in the world were as they should be.

The rest of the family also enjoyed the lazy summer days that led up to Christmas too. Ben and his friends dropped in for cold drinks before taking off on their bikes again to collect insects in a jar or ride down to the marina for an ice cream. Bill was never too far behind. After working on his calculator, he always found an excuse to go to the shops so he could drop by the boat marina where his sailing buddies had moored their boats.

Before too long I had to prepare for Christmas and drag myself around the stores for gifts and food supplies. After one of his excursions, Bill brought home a whole box of Queensland mangoes, which made a wonderful companion to the cold turkey and salad that I had planned for Christmas dinner.

Christmas morning brought the usual excitement, and Ben was in his element comparing his Christmas haul with his friends'. Even Polly perked up for a while as she waited to wolf down the turkey leftovers, before returning to her mat to sleep off the midday heat. And, as usual, Bill's happy Christmas Day turned into a sullen Boxing Day when he gazed wistfully into the TV to watch the Sydney to Hobart yacht race start without him.

By now, I was fed up with his hangdog look. 'Why don't you find someone to sail with and go in it next year?' I suggested.

'My dear, I might just do that', he replied.

After enjoying a quiet New Year's Eve we were all in bed well before midnight and looked forward to 1984 when we could finally meet our child in person. A week after New Year, I consulted the calendar and realised that I was overdue. When I saw Friday, 13 January coming up at the end of the next week, I knew it would be a special day, since the number 13 always came up before a life changing event.

On the following Thursday night I felt unusually alert and stayed up to watch the 10 'o clock news to find out about the tropical cyclone that was forming east of Townsville, off the Queensland coast. I was full of energy on that warm night; it was the best I'd felt for six weeks, and I suspected that my body was instinctively preparing me for the big day.

I smiled to myself when I settled into bed that night. 'You'd better buy some cigars for tomorrow', I said to Bill, as I turned out the bedside light.

'How do you know?' he asked.

'I just know'.

We didn't have long to wait; my labour started a few hours later, just minutes after midnight on the 13th. Instinctively, I looked toward the sleeping figure next to me for comfort. Bill was softly snoring, and in the next room, Ben also slept soundly. On second thought, I decided on a shower, and afterward I sat outside on the patio chair to enjoy the warm night and gaze at the stars and the gorgeous half moon in the sky. I welcomed the regular contractions and had no reason to believe that this labour would be any different from the previous one, when Ben was born after only four hours, from start to finish. Again, I had a feeling of oneness with all things, and felt happy and content in the simplicity of home and family. It was the same feeling that had come over me on the hilltop at Landfall, the day I had discovered my pregnancy. This time, I also felt a sense of destiny, as though a greater purpose was unfolding.

Despite being in labour, I felt quite content with no desire to wake Bill at this early stage. After a while, I grabbed a couple of juicy nectarines and settled back into bed next to my snoring husband. Staring into the darkness calmed me during the contractions, and strangely, time seemed to pass quickly, sitting up in bed all night.

That was to suddenly change at sunrise, which marked the end of the easily handled first stage of labour. I felt my body changing gears as the contractions became forceful and almost overwhelmingly painful, and I immediately recognized this as second stage labour, which means the birth is close at hand. After seven hours of first stage, it was now all systems go and I woke Bill up and told him I needed to get to the hospital quickly. He stumbled out of bed, only half awake, and foraged around the closet for something to wear. Ben was now up, after being woken by the commotion. In those days children were not usually permitted in the delivery room; and in between strong contractions, I reminded Ben to fix himself a good breakfast and told him that Bill would be back from the hospital soon with some news. In those days parents didn't routinely know the gender of their baby either, and I kissed Ben goodbye, knowing that he was hoping for a brother - even one 10 years younger would do.

After a 15 minute dash to the local hospital, Bill left me standing in the middle of the ER waiting room, in heavy labour, before disappearing again to park the car. For a few minutes, people in the waiting room stared at me until the ER staff completed my paperwork and rushed me upstairs to the delivery suite. Once there, the midwife confirmed that I was fully dilated and well into second stage labour, and phoned my obstetrician at home to let him know that he had less than 30 minutes to arrive. He lived close to the hospital and arrived 15 minutes later, gowned and gloved, but despite forceful contractions I made no further progress toward delivery in the first hour.

Before my doctor left to attend other patients, he asked the midwives to keep him updated. Every half hour he returned to check on progress, but when the second hour passed and then the third hour, it was obvious that something had gone wrong. I had been in labour far too long with a second stage that had lasted six times longer than normal, and was now in bad need of some serious pain relief. When the doctor checked me again he looked concerned, since I had developed an arrested second stage labour, a condition that required urgent intervention. Meanwhile, the midwife monitored the baby's heartbeat and announced that the baby was in distress. This got my attention, even through the pain. From my training, I knew this meant the obstetrician had to take quick action. In fact, he was at that moment trying to decide between an emergency caesarian and an emergency forceps delivery. Caesarian deliveries are typically safe when done electively or in early labour, but they were risky when done as an emergency after a prolonged second stage labour (McKelvey 2010). On the other hand, a low forceps delivery usually posed no problem in skilled hands, but my situation required a mid-cavity forceps application where the baby is rotated in a risky maneuver that requires a great deal of skill to execute. Both procedures required an epidural, and I was hugely relieved to hear the anaesthetist was on his way. Meanwhile, the delivery suite had suddenly become busy with midwives and nurses who were setting up equipment and opening sterile bundles for the procedure.

While the epidural was, mercifully, starting to work, my obstetrician explained what he intended to do and why. He told me that for some reason my uterine muscles were not able to function well enough to complete the delivery and that the baby was in distress and needed to be born immediately. Since it was important to act urgently, he chose to intervene with a forceps delivery. By then, I had entrusted myself and my unborn entirely into his capable hands, but I did briefly wonder why, since the poisoning six weeks previously, none of my muscles had worked normally.

Bill sat by my side looking stunned by the sudden activity in the delivery room. The midwife reminded him to hold my hand during the procedure. I wasn't sure who was comforting whom.

At 11 am, my obstetrician dragged our son into the world using a skilled forceps maneuver that is rarely performed nowadays. My only physical side effect from the traumatic delivery was a large tear, but our son's difficulties were just beginning. After birth, he didn't breathe and remained a dusky gray colour. The obstetrician handed him to the midwife, who rushed him across to the resuscitation table to work on him, while the doctor stayed with me to control the bleeding and suture the tear.

After a few uncomfortable minutes, the midwife had resuscitated our newborn son and we heard his first feeble cry. Not long after we breathed a sigh of relief, the baby became irritable, rigid and easily startled. His cry sounded more like a small wounded animal in distress, which was a typical sign of cerebral irritation in newborns exposed to organo-phosphate pesticides in utero, but no one diagnosed this at the time (Young 2005).

After weighing our son, the midwife wrapped him in a blanket and handed him to me. I immediately forgot about everything and melted into my newborn while Bill, too, was speechless and overcome by parental love. He stayed a while longer before leaving for home to break the news to Ben. We would call our baby boy, Justin.

Since I couldn't move my legs after the epidural, the midwife wheeled me to the ward on a bed, and placed my newborn in a crib beside me. For a while, Justin slept off his traumatic birth experience and I had a chance to gaze at him. Apart from deep bruising on the right side of his scalp from the forceps blades, he was a beautiful looking child and a clone of his father.

When the daily paper arrived, I found it impossible to focus on any other news apart from my own, but I could hardly miss the headlines that screamed: `CYCLONE GRACE'. The nurse had just brought in my roommate and settled her into the bed across from me with her tiny daughter in a crib.

`Grace is too nice a name for a cyclone', I said, trying to make small talk.

`Yes it is', she agreed. 'I like it so much that I've called my daughter Grace'.

We both laughed, but little did I know that Justin would sweep through my life, much like a cyclone, and set me onto the path toward fulfilling a personal destiny that was much different from the one I had imagined at the time.

While I was enjoying a well earned cup of tea at the hospital, Bill returned home to break the news to Ben, before phoning around to tell his sailing mates. Ben told me later that Bill had flown up the steep driveway and taken the front steps two at a time; it was the happiest he'd ever seen him. Bill's friends came by to congratulate him and a few dropped into the hospital to bring me flowers and wet the baby's head with champagne.

Despite all the excitement, the early days in hospital were enough to foreshadow that Justin wasn't going to be an average baby. He had an irritable cry, was easily startled and refused to settle for more than an hour at a time. I wondered how I would manage him at home and still take care of the household.

Five days later, I was still in a hormonal haze when Bill brought me home with Justin. Bill had abandoned his calculator for a while to revel in fatherhood; I came home to a tidy house and a cup of tea. Ben took a few days off from hanging out with his friends to bond with his baby brother. Watching Bill and Ben with Justin made me feel that my deepest longing for a family was complete, and I was ecstatic. Even Polly perked up a bit and lay near the nursery doorway as if to protect the newcomer.

The baby set the rhythm from day one and kept me busy feeding, changing, and trying to find time to do the cleaning and laundry. But despite my best efforts, tiny Justin was irritable and hardly slept, and soon the household was in chaos, and everyone was sleep deprived. Despite this, Ben handled his brother with the patience and maturity of an understanding adult, but even he couldn't settle his little brother down for long. Bill would then take over and jiggle the baby on his knee until I thought Justin would throw up – he didn't, but it didn't settle him either. I tried most tricks I could think of, including taking the baby for walks in the pram. Sometimes that did the job for a while, but usually nothing but a large feed or total exhaustion would silence the newcomer into a comatose sleep for an hour or two. Meanwhile, Bill returned to his calculator after moving his office to a quieter spot downstairs, next to the garage.

When Justin was ten days old, Bill reminded me that we needed to return to the bush in time for the start of Ben's new school year. Bill had another bit of news too. During all the baby excitement, he had kept a weather eye out for the property and Justin had brought us good luck; a soaking rain had fallen on the Western Tablelands. The drought was breaking, and it couldn't have been better news, but I was worried about going back to the bush with a baby that seemed so unsettled and now sprouted a rash at the slightest irritation. Before I went anywhere, I decided to take Justin to a paediatrician for a thorough checkup.

I encountered my first unexpected hurdle when I tried asking my GP for a referral. He seemed to take it personally when I still wanted to see a paediatrician after he'd glanced at Justin from behind his desk and declared that he was a perfectly healthy infant.

`If you insist,' he sighed, 'but I really don't think it's necessary'. The doctor showed little interest in the pesticide exposure during my pregnancy. Finally, I walked out of his office with a screaming baby, a referral and a feeling that I'd been evaluated as an overanxious mother.

A few days later, I took Justin to a paediatrician who checked his hips, tummy and head circumference. None of these routine tests are designed to diagnose the problems that I'd brought Justin in with. The predictable result of the specialist's limited examination showed there wasn't anything wrong with the baby that a 'relaxed mother couldn't deal with'.

`Babies pick up your mood', said the paediatrician. 'If you're stressed, he gets stressed too', she added, with a condescending smile.

I completely agreed with that notion, but in this case I felt there was something else wrong. Like the GP, this specialist did not consult her Merck Manual or refer to existing medical studies to assess the effects of OP poisoning in newborns. However, Dr Virginia Rauh's research explains why Justin suffered from cerebral irritation. She had studied 40 children who had been exposed to organo-phosphate pesticides before birth, and after brain scanning the children, she lists the damage done to various sectors of their developing brain:

'...the consequences of early life organo-phosphate exposure indicate subtle changes in cortical thickness and neuronal and glial cell proportions in the septal nucleus, striatum, somatosensory cortex, and hippocampus – brain regions that sub serve mood, behavior, and cognition. These studies have shown effects on glial cell numbers, changes in number and types of neurons in all four regions, as well as delayed alterations in hippocampal cell number accompanied by perikaryal swelling and excessive astrocytic processes, likely representing early stages of astrogliosis, the glial scarring that occurs in

response to cellular injury – all at a dose too low for signs of systemic toxicity'. (Rauh 2012)

Rauh's landmark study shows how acute OP chemical poisoning causes pre-natal brain injury and damages the portion of a child's brain that regulates mood, behavior and learning - thus explaining Justin's symptoms at the time, and for many years to come (Rauh 2012). In fact, it was miraculous that Justin had survived at all, since I discovered the chilling reality years later, that the damaged children in the study were exposed to doses far lower than the acute poisoning that Justin and I had experienced. Unfortunately for us, expert help was still many years away.

Meanwhile, I paid the paediatrician's substantial bill and was none the wiser after the consultation. Instead, I left the specialist's surgery with a very sick baby and doubts about my own judgment and instincts. Since I was just out of hospital and still in bad shape, I half expected a doctor to advise me to rest and recuperate from the difficult birth for at least another week before going bush. But since that didn't happen, I tried putting my concerns aside and started packing. This trip would be in separate cars; Bill would take the luggage and the dog, while I took the laundry and the kids.

Toxic Rain

As ye sow, so shall ye reap.

— *Galatians 6:7*

It was nearly dark when we arrived at Landfall, and Ben was weary. Apart from giving Ben the usual carsickness over the mountains, this drive took hours longer, with numerous stops to feed and comfort Justin. Now the baby had fallen asleep in his basket and all was quiet at last.

My headlights caught Ben swinging open the front gate while standing in water puddle; a sight I hadn't seen for two years. The 2 km drive down our private road to the homestead took us past wet paddocks, billabongs brimming with water, and a carpet of succulent new green growth spreading over the pastures. The drought breaking rain had come and brought new life with it.

The lights were on at the homestead; Bill must have arrived hours before with Polly, and no doubt wondered where we were. On the hill, the workman's cottage was also lit up, with Allen's Ute parked in the drive. It was reassuring to have another family living on the property, especially with a new baby.

That evening, we reunited over a can of beans that Bill had found in the pantry, while, thankfully, Justin stayed asleep.

'What did the doctor say about Justin?' asked Bill.

When I told him about the clean bill of health, Bill looked relieved, but from the look on my face, he could tell I wasn't buying it.

For the rest of the evening, the three of us were lost in our own thoughts. Ben was sorry to leave his friends at Bayview and not looking forward to the end of his summer holiday. I was wondering what life on the property would be like now the drought had broken. The one thing I knew for sure was that I would be unable to rest and settle in with the new baby since I would need to go to town, the next day, to restock the fridge and pantry after our time away. Bill wouldn't be able to help me much, since he was planning a huge salvage operation after the drought's devastation, and the work would have to start in the morning.

The next day, Allen and Darren turned up for work, and during their usual meeting in the homestead driveway, Bill gave them a long list of chores that would keep them busy for weeks. Afterwards, both men drove off in Allen's Ute, while Bill jumped into the Rover, this time without me, to personally survey the damage.

Meanwhile, the new baby attracted the usual female curiosity. When Barbara dropped in especially to see Justin, she delighted in cuddling him, but her cooing didn't stop him crying.

'I see you've got your hands full', said Barbara, rocking him ever more furiously.

'This baby could cure your cluckiness', I answered over the noise.

'Don't worry', said Barbara sympathetically as she handed him back to me. 'I know it's your turn to drive the kids to the bus-stop, but I'll keep going until you get back on your feet'. For once, I swallowed my pride and accepted her offer. If today was any indication, I would need all the help I could get, since I still hadn't made it to town for the groceries.

Justin had just gone down for a short nap when Bill arrived back at noon. He threw his hat on the table, slumped in his chair and waited for the lunch that I hadn't made yet.

'That bad, huh?' I asked cautiously. 'I thought you've wanted rain for two years'.

'Of course', said Bill. 'But the grain silo's empty, the hay-shed's bare, there's no Lucerne in the 20 hectare fattening pasture. It's like starting all over again'.

'What's the stock like?' I asked from the kitchen, while searching the pantry for something to cook for lunch.

Bill came into the kitchen to put the kettle on. 'I've had to sell so many sheep to spare the paddocks that it'll take months to get the numbers up again'.

'What about the cattle?'

Bill returned to the table with his mug of coffee. 'We'll have to bring the herds back from Victoria and from the North Coast', he said. 'The agisted cows calved while they were away and we have twice as many to bring home', he added.

'That's good!' I shouted and tried the half glass full approach to cheer Bill up.

'It would be, if we didn't have another problem', he replied.

'What's that?'

'Come out with me and I'll show you'.

Over a canned soup lunch, I told Bill that I couldn't imagine how I could accompany him on his usual rounds with a newborn baby, but promised to give it some thought, if I could find a safe way to transport Justin. I guessed I wasn't the first grazier's wife to take her newborn out to inspect the paddocks.

The new school year started the following week and Ben left each morning to catch a ride with Barbara, while I stayed home to give Justin his morning bath and feed.

Meanwhile, Bill sat at his coffee table in the front room and worked his calculator. He was worried about our stock coming back and tried to calculate if our pastures had the carrying capacity for the stock numbers. Despite all hardships, our stock had fared well in the drought. The other graziers in the district, without cash reserves or large holdings, had no choice but to sell their stock to the abattoirs once the fodder was eaten, while those who'd hung on to stock for too long in hopes it would rain had ended up digging lime pits for their starving animals. As always, battlers with bank loans always fared worse than the large agricultural holdings or consortiums. Bill and I were somewhere in the middle. We had no bank loan over the property, and a second income from investments. We'd survived this time only because Bill was an astute businessman and a good drought manager. We could afford to top up our silos and hay-sheds, which enabled us to keep the best breeding stock to start over with when the drought broke. But Bill and I both knew that another drought anytime soon would wipe us out too.

Now that our cattle were on their way back, Bill scouted out the pastures. I was able to join Bill after I'd made a sling that allowed me to carry the baby close to my chest, where he seemed happiest. From the Rover, I watched Bill survey the paddocks, his eyes darting here and there. He never missed a new clover leaf or a fresh blade of grass. I knew the fate of our stock rested heavy on his shoulders, and it wasn't long before he spotted the next problem.

'Take a look at that', said Bill pointing to a thick carpet of variegated thistle that had crept across the countryside. Weeds always proliferated after a drought and this variety of thistle grew so quickly that if left unchecked, it spread over the entire property and choke out any useful grasses. This meant our stock would go hungry in knee-high thistle, and it would also contaminate our next wool clip with vegetable matter. Other noxious weeds, including blackberry bushes and cacti were also spreading. Bill explained that we had to get rid of these weeds and improve the pastures all over again if we wanted a going concern the following year. That meant weeding, reseeding, and restocking thousands of acres, and it represented a huge investment of time and money.

Bill never gambled, not even on the pokies, but he realised that for the next couple of years we would be gambling our life savings on decent rainfall as we tried to recover from this record breaking drought.

The immense cost of rehabilitating the property was just beginning to sink in when Bill discovered yet another problem. This time a plague of insect pests, the red-legged earth mite, had spread throughout the Lucerne paddock, and it wouldn't be long before the newly planted sorghum and oats would be infested. He didn't tell me at the time, but this meant only one thing: barrels of chemical pesticides and herbicides.

Many years later, I discovered that both pests and weeds flourish almost exclusively in the kind of mineral depleted soil that is typically created by a long drought. Much of the topsoil is lost, along with the humus, while the struggling drought-affected plants extract what's left of the soil's natural mineral content. While we were trying to solve our pest and weed problem, a French agronomist in Europe was compiling a remarkable scientific paper based on his new trophobiosis theory (Chaboussou 1985). Francis Chaboussou discovered that a plant grown in soil containing sufficient minerals is too healthy to attract pests. In his field work he observed what organic gardeners already knew; a pest starves on a healthy plant. However, since the 'green revolution', chemically-based agriculture has focused on killing the pest with pesticides, instead of adding minerals to the soil. John Fenner of the Australian National University School of Environment & Society has reviewed Chaboussou's research and writes:

Pests shun healthy plants. Pesticides weaken plants. Weakened plants open the door to pests and disease. Hence pesticides precipitate pest attack and disease susceptibility, and thus they induce a cycle of further pesticide use.

At the time Bill and I did not know about Chaboussou's work, which has since inspired the development of several soil (and animal) mineral supplements that can be applied to mineral depleted soil to restore its vitality and reduce or eliminate pesticide and herbicide use[1]. This mineral solution also enables the healthy plants to pass on enough minerals to the animals that graze on them, and up the food chain to humans. However, this natural approach is not popular with proponents of chemicals and synthetic fertilisers, and not promoted by governments that traditionally support multi-national industries. On this topic, Fenner notes:

It is certainly the case that agribusiness continues its focus, not on the health of the crop, but rather on the demise of the pest, and so continues to develop novel pesticides, genetically modified organisms that

---

[1] http://www.tnn.com.au/

produce pesticides or can withstand heavy pesticide dosages, and most recently the coupling of nanotechnology and pesticides.

Justin was only a few weeks old when the herbicide 2, 4-D and the organo-phosphate pesticide, Dimethoate, rained down from the sky. Bill had made a phone call to the Ag-sprayer, and the crop duster just appeared in the sky, without warning, and dropped its payload. Bill was doing what he was required to do as a landowner. Sooner or later, the weed inspector from the Pastures Protection Board would come around and order us to chemically spray the property, or cop a fine. The Board gave out no information about natural or non-toxic options, either then or now.

The spray had a strong chemical smell and left fumes lingering in the air for days. I felt nauseated after breathing it and developed a severe headache and nausea.

Ben was at school during the spraying and showed no immediate ill effects, but Justin was distressed and unwell. After exhausting crying jags, he slept a short while before waking up again, sweaty and easily startled.

At the time, I didn't connect the dots that the spray also ended up on the roof and ran off into the water tanks, only to affect all of us, for as long as we used the water. And sure enough, a few days later, Ben came down with what appeared to be a bad case of the flu. It was rare for him to be sick; he'd hardly had a day off school for three years, but this bout sent him to bed, and it wasn't even the winter flu season yet.

While the kids were both unwell, I had my first ever bout of asthma. After two suffocating hours, Bill had to dash to the chemist in town and get me an asthma inhaler. That night, I begged Bill never to subject us to aerial spraying again; I was sure the kids and I were sick because of it.

Bill's turn came when he walked through the paddocks too soon after they were sprayed to see how effective the eradication had been. The thistle was browning off, but so were the peppercorn trees that gave shade in the paddocks. Bill walked through the Lucerne paddock while strafing his working pants at thigh and groin level. The mites were dying off by the millions, as were the ladybirds that came to feast on them. Days later, the air was still thick with chemical vapors and killing off natural predators again.

I threw Bill's working pants, impregnated with herbicide and pesticide, into the dirty laundry basket, including the baby's clothes (unaware that these chemicals cling to fabrics for weeks or longer). The laundry piled up in the basket until I had time to take the washing to the Laundromat in town, since Bill hadn't replaced the old Hoover twin-tub, in case the rodents made a meal of it again.

A week later, Bill complained of an angry red rash around his groin and when he started to complain of dizziness, I convinced him to see our local doctor in town. He came out of the doctor's surgery with a prescription for anti nausea tablets and a steroid cream for the rash. After one tablet made him feel woozy, he threw the pills away and only used the cream, which eventually cleared up the rash, but the giddiness remained for several years more.

No sooner had the spraying finished, when the cattle transports started arriving, one by one, over the next three days. Three hundred Herefords and their two hundred calves stepped off the trucks, caked in dung after traveling such long distances. The dipping and drenching began again since those animals couldn't be let in with the stock at home before they'd been treated for ticks and parasites they might have acquired while away.

This time we were better staffed, and Allen did the job with Darren and an extra casual farmhand. The three men worked a hard and fast week so Allen could return to his cottage for a couple of beers, at 5 pm sharp. By weekend, Allen had accumulated a small mountain of empty beer cans next to the verandah, and I hadn't seen Barbara and the kids around all weekend. I needed to remind her that it was my week to do the bus-run. When I went up to the cottage to let her know, I found Allen sitting on the porch glassy eyed with a beer in his hand. He told me that I'd better count on doing the run permanently from now on, since Barbara and the kids were gone.

'When will she be back?' I asked, innocently assuming that she had taken the kids on a holiday.

'She won't be', replied Allen, taking another swig from the can.

Confused, I returned to the homestead to tell Bill about it.

'Those blokes are all on the turps when they knock off', explained Bill. 'There isn't one any different'.

'Do you think his wife left him?' I asked.

Bill shrugged his shoulders. 'Who knows?'

Just as the workload settled back into a manageable routine for Allen and Darren, I sensed Bill's preoccupation about something. It wasn't long before I discovered two things about Bill: his first love was the sea and nothing could keep him away from sailing for very long. I already knew that once my husband made a decision, he wasted no time in carrying it out, and Bill was planning to go sailing.

A few days later, I left Bill in charge of Justin while I dropped Ben off at the bus-stop. When I returned, Bill had already packed up his car and was waiting for me to relieve him. As soon as I got out of the Honda, Bill plastered a wet kiss on my cheek.

'See you in a couple of weeks', he said. 'I left a cheque on the dining-room table', he added, as he threw his sailing waterproofs into the car. Through the dusty windscreen, his face shone with adventure. From the driveway, Polly and I both watched as Bill's diesel Mercedes disappeared from view in a cloud of blue smoke. I didn't feel much like smiling; the two aspirin that were supposed to get rid of my headache were stuck half way down, and left me with heart burn. Lately I was never without a headache. For a moment, I felt disappointed that Bill was taking off, just when we had a newborn baby and I needed him most, but another part of me yearned to be on my own for a while. I headed back inside and wondered what I would do while he was gone and soon decided that after bathing and feeding Justin I would plan an interesting day, before Ben was due home from school.

Lately, I felt tired even after a night's sleep, and thought that I had become so unfit that a short hike around the paddocks, closest to the homestead, would do me good and maybe shift my constant low grade headache too. Plenty of household chores waited for me, but Bill's absence and the autumn sunshine made the idea of a walk irresistible.

Walkabout

Walkabout: a rite of passage, a journey in the wilderness.

- Wikipedia

I stood in the driveway, with Justin in a sling, trying to decide which bush track to take. Polly looked up and wagged her stumpy tail, before deciding to come with me. I was going 'walkabout', originally an Australian Aboriginal rite of passage in the wilderness, but in recent years the term meant a bush walk with no specific destination. With Bill gone, I was in no hurry.

I glanced toward the fuel shed, trying to decide whether to take the road that led up the hill overlooking the homestead, when an astonishing sight greeted me. Next to the shed, on the forked branch of a giant ghost gum, a huge goanna lay sunning himself. I approached the enormous lizard carefully until I was close enough to see the lacey detail on his yellow and brown flecked hide. Polly followed behind me and sat down looking mesmerised. The reptile must have been close to two meters long and lay completely motionless, without the slightest breathing movement. He could have passed for the fossilized remains of a prehistoric creature, except from under his half-closed lids stared a primal and intelligent pair of eyes. His soft earthy colours perfectly matched the surroundings. The Australian Aboriginal tradition valued the goanna as the keeper of secrets and the guardian of knowledge. I wondered what knowledge this primeval creature was trying to convey to me.

As the goanna and I stood facing each other, I felt on the verge of discovering some of the signs that are spoken of in the Aboriginal culture, which teaches that signs can be anywhere in Nature and they always conveyed profound meaning. From the beginning, I'd sensed the land was steeped in mystery. Later I discovered that the Wiradjuri Aboriginal people had lived in this region in tribal groups, before Europeans came. According to DNA findings, Aboriginal origins go back 400 000 years, through four ice ages, to the very beginning of man as Homo Sapien (Lawler 1991). They lived here long before the Stone Age people populated Europe, and in a continuous lineage up to the present time. Their lifestyle represents the Natural way that man was intended to live in a balanced and harmonious relationship with Nature and as part of the web of life. Their myths describe the beginning of creation itself, and to understand them is to look into the very origins of our species.

Now I stood before the keeper of Indigenous secrets and was so mesmerised by its presence that I experienced a strange flash of insight that is impossible to convey in words, but it was as if I were shown a blueprint of my life. This led to a flashback to my 6 year old self, standing on the porch of my parents' house in Michigan, where I was shown a future that had to do with health and healing. That had certainly become a reality when I pursued a career in healthcare, but now I was shown another glimpse into my future; an image of myself standing in front of thousands of people. At that time, I couldn't have understood the prophetic significance of this.

Instead, I felt slightly dazed but calm, as evidently did Justin, who was sleeping soundly, cradled in the sling. Polly had also fallen asleep at my feet.

As I was about to snap out of the dreamy mood and start my walk, I noticed Allen's red Ute in his driveway. It seemed strange that he was at home this time of the morning and not out working with Darren. I thought he was having his morning smoko, and dismissed it, before heading off in the opposite direction. When I glanced back one last time, the goanna was still guarding our homestead with the unblinking stare of a watchdog.

With Justin fast asleep, I set out past the house dam and saw none of the usual ducks and native birds feeding on tadpoles and insects among the pampas grass. Still wondering why the area was devoid of life, I continued up the track to the Fossicker's paddock, one of Ben's favourite places to go looking for quartz that contained flecks of gold. During the 1881 gold rush, prospectors had staked claims around this paddock, but the richer deposits around Eurunderee creek yielded more gold, and our paddock was soon forgotten by all but a few old timers who came to fossick around on it from time to time. Even now, only bracken ferns grew among the tailings left by the prospectors a hundred years earlier. That bracken was totally browned off, with a faint chemical smell still lingering in the air. Now I knew why the ducks and insects weren't in the dam - and why the plants had wilted; because of the aerial spray two weeks earlier.

Next to Fossicker's was the upper oat paddock that shared a boundary with the national park. Polly sniffed the air and alerted me to a group of kangaroos grazing on the oat stubble. When I stopped to watch them enjoying their four-star dining experience, they noticed me and lifted their heads up. Looking startled, they balanced on their tails like a tripod, before clearing the boundary fence in one giant leap, with their pouches full. The big grey male lingered until all the females had gone to high ground, before looking at me as if to say: 'Don't mess with them, or you've got me to answer to'. His front paws were braced, like a boxer, ready for action.

'Don't worry mate', I said out loud. 'I'm just another female with a joey'.

As though he understood, the big grey kangaroo balanced on his tail, before jumping the fence to join the females.

By then, I was beginning to feel Justin's weight, and before too long I found a shady place under a eucalypt tree, near a dam, where I could rest and feed Justin, with Polly snoozing beside me.

After the rest stop, I decided to head back to the homestead via the old dairy and check on the two ferreters who had asked permission to stay there overnight to catch some rabbits. On my way, I noticed a large pile of wool on the ground, in the distance, indicating that a sheep was down. Polly's nose had already picked up the scent of death, but I still thought the animal was alive and approached closer to see if I could help. It was then that I noticed not one, but two sheep were lying on the ground lifeless. In the bush, a fly trail always points to the cause of an animal's death, and the flies had already laid maggots into the circular head wounds of the two dead sheep. I wondered what was going on; there were times when the farmhands had to euthanize sickly or injured animals with a single shot from a .22 rifle, but two at a time seemed unlikely, and these sheep had died suddenly, in the prime of health. I toyed with the idea of letting Allen know, but something didn't feel right about that either.

After that grisly scene, I approached the old dairy more cautiously than usual. Even from a distance, I saw that the two rabbit ferreters, Reggie and Harold, were in residence; their rusty Holden Monaro was parked outside, with a trailer full of caged ferrets. They usually drove across from Western Sydney, one weekend a month, and seemed so happy camping in the cottage, among rodent droppings and bits of old lino, that you would have thought they were staying at the Ritz. After a weekend of ferreting out rabbits from warrens, drinking Foster's lager and not showering, the pair smelled worse than their ferrets. If the old boys neglected themselves, their ferrets, by contrast, were kept in top racing condition. I smiled as I recalled the day that Bill and I had first met the pair, when Reggie held up one beady-eyed creature by the scruff and announced, with the respect usually reserved for a Melbourne Cup winner: 'This is the red-eye doe'.

Not to be outdone, Harold held up the other ferret. 'This is the black-eye buck. The piss de resistance!'

They explained that the ferrets were trained to scurry down rabbit holes and flush out entire rabbit colonies, and you had to be quick to net the panicked rabbits as they escaped from their rabbit holes.

Aside from Bill's obvious amusement when the pair came on safari, Bill tolerated ferreting as a form of biological pest control, but he still thought of it as more of a joke than a way of overcoming the rabbit problem. We'd poisoned most of the rabbits with 10-80 the year before, and the two good old boys were welcome to what was left.

As I approached the old dairy, Polly went over to greet the boys, who were just loading up their rabbit catch and securing the tarpaulin over the trailer for the trip home to Blacktown, while the ferrets were asleep in their cages, exhausted.

'Hi there, Missus!' shouted Reggie, as he spotted me with Justin. 'Want a bunny for the little fella's dinner?'

'No thanks, he's not old enough for rabbit stew yet', I said, as I leaned on the fencepost and watched them for a while. 'So, how did you boys get away from the missus this time?' I asked. It was a running joke between us by now.

'Same as usual', said Harold, with a toothless grin. 'When the missus come in the front door, we git out the back, quick smart!'

I laughed with them, as usual, but the mood turned more serious when I asked them whether they carried a rifle. With a wounded look they explained that they owned no guns and didn't consider it a sporting implement in any case. That was the whole point of ferreting; to give the game a chance to outwit the hunter. I believed them. After all, I had never seen them with a gun. The only weapon they carried was a trailer full of ferrets.

While trudging back home, I was starting to wonder if I would make it. My limbs felt leaden, and I was on the verge of another asthma attack and would have to hurry back to the homestead for my puffer medication. Just as I tackled the last hill before the house paddock, I noticed that a cow was down. With Justin in tow, I worried that I wouldn't be able to help the animal. But I needn't have fretted; when the wind changed, I didn't need Polly's nose to tell me that it was far too late to help. Closer up, I saw the Hereford cow lying dead with the same circular wound as the sheep I had seen earlier, but this animal had been shot through the head *and* the chest. Obviously there was a lot going on that neither Bill nor I knew about. Harry and Warren used to let us know when any stock had to be destroyed, so we didn't suspect foul play, but these animals had been in perfect condition when they'd died. Now it seemed that someone was deliberately shooting our stock, but who - and why?

Polly and I were both relieved to arrive home, and while Polly slumped on her mat, I headed straight for the medicine cabinet and took two deep puffs from the Ventolin inhaler. While waiting for it to work, I felt like collapsing into bed, but Justin needed another feed and it was time to collect Ben from the bus.

A few minutes later, when my airways were open again, I felt well enough to drive. When I picked up the car keys from the table, Bill's cheque was lying there, and I saw that he had written it out for only $100. This meant that we only had $50 per week to live on, and even in 1984, this was not going to meet our needs for the next two weeks, especially with a new-born baby. I had no access to investments or cheque accounts, and no amount of discussion had changed Bill's ways with money. Now he was at sea, completely out of contact, and we would somehow have to make do.

On my way to the bus-stop that day, I suddenly felt unprotected and isolated out in the middle of nowhere, on my own, with two children. Our property was twice the size of Gibraltar, and I had no meat or vegetables in the fridge or much petrol in the tank. The situation seemed ridiculous, considering we owned more arable land than some countries in the world, yet I was completely dependent on Bill's largess for our living needs.

That night I was too tired and ill to eat dinner, but somehow managed to feed the kids early, hoping to snatch a few hours of sleep before Justin woke in the night. It had been a long day since this morning when I'd stopped to stare at the goanna in the ghost gum. In bed that night, my mind reeled with the day's strange events, before falling into an exhausted sleep, with Justin in the bassinet beside me.

Justin woke me up a few hours later and I quickly reached over to bring him into bed with me, where I would change and feed him, before his crying woke Ben up. That night, I had an eerie feeling and didn't want to turn on the bedroom lights with only a flimsy lace curtain barely covering the window. Bill maintained that, so far out in the bush, there wasn't much point hanging drapes on windows, especially with the workman's cottage on the opposite side of the house, and not in view of the bedroom. Now I wished he had hung them, so I could turn on the lights without fear of being seen by someone outside the house.

I sat up in bed and fed Justin in the dark while peering through the sheer curtains to the moonlit hills outside. The digital clock on the dresser showed it was 1.35 am, and I listened intently to the usual creaks and groans of the bush house. Just as I convinced myself that, with Bill gone, my imagination was running wild, I spotted a light in the distance, through the window, and I locked my eyes on and followed it until it divided in two. It looked like a vehicle's headlights were trailing up the winding road, until finally disappearing behind the granite ridge to the north. I wondered what business anyone had on our property this time of night, and my annoyance provoked me to get out of bed to investigate. I was on my own though, since Polly was no watchdog, and never had been. I wrapped Justin in his fleecy blanket and tucked him securely into my warm bed. Miraculously, he made no fuss when I got up. When

I felt my way through the dark house I noticed Ben's bedroom door was ajar, and he seemed peacefully asleep. Quietly, I moved ahead through the winter room where the long burlap curtain separated the derelict old cottage at the rear, from the homestead. It reminded me that the creepy rear section, where the previous owner had taken his own life, had several unlocked external doors that gave access from the outside directly into the homestead. No doubt, I would have to go out there at some stage, but I certainly wasn't going out there right now, in the pitch dark, trying to lock doors that hadn't been used for years. Instead, I kept moving, into the dining area, past the wicker chairs and the long row of windows that overlooked the hills where I'd seen the lights.

Next, I searched around the dark living room for the front door. The doorknob turned easily, but the lock was rusted and couldn't be locked. From there, I continued into the kitchen and peered through the window, where I could see the corner of Allen's cottage. It stood in darkness, but I couldn't tell if his Ute was there and wasn't game to go outside and check it out from our front gate. When there seemed nothing more I could do, I felt my way back to the bedroom in the dark, with every nerve on alert. Back in bed, I cradled Justin and spent the night staring at the glowing digital alarm, and counting the minutes until I fell into an exhausted sleep around 4 am.

At sunrise, Justin woke up again and the world looked brighter, but I was exhausted and every muscle ached from the hike I'd taken yesterday. There was no rest from the daily routine though; Justin needed his bath and feed, Ben had to go to school and I had to drive to town to cash Bill's cheque, tank up with petrol, buy a few groceries, and hope the hundred dollars would keep us going for the next few weeks. First of all though, I phoned the police station to let them know what was happening on our property. The senior constable told me that a lot of land owners complain about people coming onto their property, but without evidence, they could not do anything about it. He advised me to get a license plate number and phone them back with that information.

Polly kept me company over the next few sunny mornings, sitting under the chestnut tree with Justin, and trying to recuperate from the body aches and pains that had set in. I realized that something had changed in my body since the poisoning and I could no longer count on my previous muscular fitness. For the first time ever, exercise did not make me fitter, but exhausted me, made me sore and feverish, and set off such a wheeze in my chest that I had to sit and rest between bouts of housework.

From my vantage point in the front yard I watched the Kookaburra family in the Chestnut tree and the Suphur-crested Cockatoos picking seeds from the lawn, but it also offered a clear view to Allen's cottage. Sure enough, there was no washing on the line, Barbara and children were nowhere to be seen and Allen was elusive. In the past week I hadn't even seen Darren, who used to report to work each morning with Allen, as long as Bill was around.

With Bill away at sea for another week, there didn't seem much more I could do, except keep an eye on the place, and get on with some horse work that was overdue. The three pure bred mares were back from stud in Victoria and two months in foal. Their older foals, two delightful purebred fillies and a colt, needed handling. For that job, I called in the local horse breaker, Joe Glover, who helped me round up the horses from the front paddock into the stockyards.

When Joe started working with the young horses, Allen turned up looking sour and asked how long we were going to be using the yards. Joe told him he would need them for the next two days to break the fillies in to halter, and after that he would float the colt over to his place and work with him there over the next week. On hearing this, Allen walked away with the look of a man who was afraid that someone would ruin his plans.

It was a pleasure to watch Joe working the horses, over the next two days. With Justin in his basket under a shady tree, I helped him handle the fillies until they accepted a headstall and halter and allowed us to lead them around the yards. After just a few days of work though, I felt weak again and was glad to see the end of the job.

Meanwhile, between the hard work, sleepless nights and worry about not having enough money to feed us, my milk supply dried up and I could no longer feed 6 week old Justin. The next day I had to drive to the town pharmacy and purchase infant formula, baby bottles and sterilizing solution. It cost me almost all the money I had left out of the $100 cheque, but the final budget breaker turned out to be a nearly empty petrol tank from the runs to the bus stop, and trips into town. From now on, I wouldn't be driving very far until Bill returned home.

The days went by quickly enough, but the nights proved increasingly sleepless as Justin started having colic from the bottle feeds. Years later I realized that I had used unfiltered tank water to mix his feeds, and though I had boiled the water first, boiling would not have removed the pesticide residue from the aerial spraying. This only made Justin's problems worse, which now included a rash that came and went on his back and tummy.

For the next week, while I was up with Justin at night, I noticed headlights beaming from various areas around the paddocks, at all hours of the night. I still couldn't contact Bill, and didn't trust Allen enough to ask him about it, but I did feel a heavy responsibility to keep my children safe. As it turned out, I had every reason to be worried; after Allen's wife had left him for good and taken the kids with her, he was spending most of his time at the pub in town, and returning to the cottage late at night where he would drive his Ute drunkenly around the property with a variety of rifles that he owned.

Even though I didn't know exactly what Allen was up to at the time, I thought it was sensible to use the daylight hours to check out the disused rear section of the house and barricade the back doors with old furniture to make it more difficult for someone to gain access to the homestead through the back. Of course, this would pose no problem to anyone determined to enter, plus the front door still couldn't be locked at all, so for good measure, I left the number of the Mudgee police station next to the phone, in case there was a break-in. I was aware that in the event of a serious incident, the police would have taken up to an hour to arrive, since we were well outside town. The neighbors too, were miles away, and without much petrol, I wouldn't get very far if I needed to escape, with the kids, from a situation.

After two weeks, the constant tension and sleepless nights were beginning to take their toll on my nerves. Every night, after the kids were in bed, I phoned old friends in both Sydney and Adelaide. They kept me company on the phone, but beyond half-hearted reassurances, they couldn't do much to keep us any safer. Neither Bill nor I had any family members, or even distant relatives in Australia that I could call on for help. I couldn't help but realise the full implications of my position: that I was well and truly on my own.

One night I hung up the phone and cried inconsolably for hours. After that, I blew my nose and was determined to regain some courage and to stop feeling like a victim. I decided to be strong for the kids, since their wellbeing and safety was entirely in my hands. That meant that I would do whatever was necessary to keep myself and the kids safe.

On Friday afternoon, Ben arrived home from school, excited that it was the beginning of the Easter break. I suggested that he might like to catch Bridie, his Welsh Pony, and put her in the stable overnight so he could ride her over the weekend.

That meant using a small amount of oats to lure the pony close enough to maneuver the lead around her neck while she was eating. Sometimes the cunning little mare let Ben catch her, but pig-rooted when he tried to saddle and ride her. That afternoon, Ben had fun trying to outwit the pony until dinner time, when he came inside grubby, but pleased with himself, and headed for the shower.

I was frying sausages when the phone rang on the coffee table that Bill had turned into his office desk. It was Bill calling from the Cruising Yacht Club in Sydney.

'Ditch'ya miss me?' he asked.

'You could say that', I answered brusquely, after spending the past two weeks without enough money to live on. But Bill didn't notice my bad mood, since he was still on a high after he'd sailed a 50 foot yacht from Hobart, across Bass Strait, up the South Coast, and into Sydney Harbour to the CYC. Tonight he would stay on board the boat at the marina, he explained, and the next morning he and a sailing mate would sail her up the coast into Pittwater. He wanted me to drive to Bayview with the kids and stay with him over the Easter break.

'Waddaya think?' he persisted.

Since Bill hadn't explained much to me before leaving on the trip, I just thought he'd gone on a sailing trip with one of his buddies who owned a yacht, and I told him that I couldn't see much point in driving all the way to Sydney to see some guy's boat. Instead, I tried explaining about the lights in the paddock at night and insisted he come home to sort it out as soon as possible.

'Oh, don't worry about that', said Bill. 'It's probably one of the boys spotlighting or Roo shooting'.

Bill's euphoria couldn't be dampened, and he used the next few minutes to convince me that I was overreacting, and that a week in Bayview would do me good.

I flared up in frustration. 'That's easy for you to say, but how do I get there without petrol?'

'There should be some in the drums in the fuel shed', he replied chirpily. 'And by the way, you should remind Allen to feed the dog for us'.

After hanging up the phone, I thought we were better off to eat dinner, since Justin had settled for the moment and Ben was freshly showered and waiting patiently at the table in his flannel pyjamas. I would check the fuel shed in the morning and decide then whether we could go to Sydney or not. Just as I was in the kitchen dishing up dinner, something caught my eye through the window. It was nearly dark, but I had a direct view to the cattle yards 500 meters down our private road. This time, I rushed outside and spotted the faint outline of a vehicle. From my garden gate, I saw a set of red brake lights flicking on and off as a large truck seemed to be maneuvering out of the stock yard parking area. I wondered whose vehicle it was and checked Allen's drive, but realized that the vehicle coming out of the yards was too big to be a Ute. Even though vehicles on our property weren't unusual, no stock movements were scheduled at the time, and that size truck had no reason to be on the property. I felt a strong urge to go down there and check it out, but was hesitant about confronting strangers who obviously had the nerve to trespass on someone else's property, whenever they pleased. I dashed across the driveway to the car shed to see if I could find a car jack or any implement to take with me for protection, but there was only the .22 rifle lying on the top shelf, and I thought the better of bringing it.

Ben had been watching me from the living room window and rushed outside, with Justin, to see what was going on. For a moment, I deliberated whether it was safer to leave the kids alone in the house or take them with me, and in a split second decision, I bundled Ben into the front seat, with the baby, and took off down the road after the vehicle. By the time we got to the yards, the truck had gained a lengthy head start and disappeared into a dust cloud as it hurtled toward our front gate. With the kids in tow, I pursued the vehicle for the next kilometer to our main entrance, only to find nothing but the front gate still swinging on its hinges after someone had made a hasty exit. There was no chance of getting a license plate number, but plenty of evidence was imprinted on the ground where the wide tyres had left marks in the soft road shoulder. Judging from the depth of the marks, it had been a fair size truck with a heavy load on board.

That night, after barricading the remaining doors that couldn't be locked, I thought about the .22 in the car shed. In those days every farm had a small caliber rifle to euthanize animals or to cull kangaroos and rabbits. In the bush I'd often gone target shooting and spotlighting for rabbits, and I wasn't a bad shot. In the days before stricter legislation, the .22 often lived somewhere close at hand in the bush – where it could be chucked into the Ute before going out spotlighting. But under these circumstances, I decided that the .22 was coming inside with me, where I found a safe place for it on top of my bedroom clothes cupboard, along with a box of ammo.

That night was as long as all the ones before, but the thought that I could defend myself against an intruder made for a much better sleep.

First thing next morning I drove my car up to the fuel shed, where I hadn't been in three years, but soon I remembered Bill telling me of a hand pump that was usually attached to a 44 gallon drum. Luckily, there was liquid in one of the drums, and after making sure it was petrol and not diesel, I was able to fill the tank of my Honda wagon with enough fuel to get me to Sydney. It was the fist time I'd noticed becoming dizzy around petrol fumes and I couldn't wait to get out of that corrugated iron fume closet. With that job done, I stood outside in the fresh air for a minute, while working up the nerve to go across to Allen's cottage. It was bad enough that I had to let him know I was leaving for Sydney, since one of his jobs was to feed Polly while Bill and I were both away, but I refused to give him the benefit of knowing how long we would be gone.

I felt nervous knocking on Allen's door. For a minute all was quiet inside, until the door slowly opened a crack and

Allen peered out with yellow tinged eyes, wearing the dirty jeans from the day before. I explained my plans through the narrow opening.

'No Worries', he said with breath smelling of alcohol and cigarettes.

On my way back to the car, I passed Allen's Utility and noticed the tarp flap open, exposing a couple of high powered rifles with telescopic sights in the rear tray. That kind of weaponry was not illegal at the time, but it was more suited to a big game hunter or a sniper, than a farm manager. That made up my mind for me. I returned to the homestead, packed up the car and put the kids inside, and then called Polly. She was coming with us, even though Bill had wanted her to stay. Soon after, I pulled out of the driveway with the kids, the dog and enough clothes to see us through the week's Easter break at Bayview. I tried not to think about what Allen would be up to while Bill and I were both gone, and I couldn't get out of there fast enough.

Betrayal

This discipline and rough treatment are a furnace
To extract the silver from the dross.
This testing purifies the gold -
By boiling the scum away.

- Rumi

After an exhausting drive from Mudgee with the kids and Polly, I found Bill relaxing on the veranda at Bayview with a glass of rum and cola and a two week beard. I tried telling him about the goings on in the bush, but he was still in no mood to hear about it. Instead, he beamed a grin as wide as the bay, and I thought he was just happy to see me, until I discovered the real reason for his good mood, when he insisted on taking us down to the marina. With baby in tow, Ben and I barely kept up with Bill as he raced ahead to the end of the dock and pointed to the first row of yachts tied to their moorings. The biggest of them was a sleek white yacht with a tall mast; on her stern in gold letters was *Encore*.

'She's a forty seven foot Kelly Peterson', said Bill reverently, 'designed by Doug Peterson. One of only 30 built'.

I looked at Bill blankly.

'Don't you want to go on board OUR boat?' Bill asked.

'Our boat?' I repeated dumbstruck.

One minute I was battling in the bush and the next minute Bill had us all in a dinghy and motoring towards a sumptuous ocean cruising yacht. I couldn't have been more stunned. After we climbed up the yacht's rear boarding ladder, Ben went down below to explore the living quarters, while Bill and I stood on the sun drenched top deck holding Justin. Together we surveyed the 47 feet of sailing elegance, right up to the bow.

'You wouldn't be dead for Quids', uttered Bill, almost reverently.

I heard Ben shouting from down below. 'Mum, this is gonna be my bedroom!' He could hardly contain himself as he explored the yacht's roomy interior.

Encore was generously appointed with double bedroom cabins fore and aft, a navigation bunk on the starboard side where Bill spent most of his time, and two marine toilets with showers. She had a gourmet galley with a double sink and a roomy saloon with a teak table that could comfortably seat eight dinner guests.

'So, I guess this will solve your Boxing Day blues', I said to Bill.

'This won't be winning any Sydney Hobart races' he laughed, 'but maybe the Sydney to Noumea race…'.

That was Bill's way of telling me that his previous boats were fast, but uncomfortable ocean racers that attracted an endless supply of young and fit guys who wanted to crew in competitive ocean races. But *Encore* was an ocean cruising yacht, designed for long distance blue water sailing or on-board living, and she didn't come with that kind of male muscle.

'What do you know about sailing?' Bill asked hopefully, looking from me to Ben.

I looked at him in disbelief. The fact that he had bought this yacht without so much as discussing it with me was now sinking in. 'Maybe you should have asked me that before!' I fired back.

Fortunately, I knew something about sailing. I told Bill that before we'd met I had regularly crewed for years on a friend's 30 ft sloop for the summer races at the Garden Island Yacht Club, Port Adelaide. Sometimes Ben came with me and even though he'd learnt how to winch in the sheets when he was only five years old, neither of us had ever handled anything like *Encore* before. But just like Bill, I'd never steered clear of any interesting challenges before and wasn't about to start now.

Bill looked me up and down to assess if I could handle the risks, before deciding that I might have what it takes, after all. At that moment, we literally placed our lives into each other's hands, and made the decision to handle the boat by ourselves. My husband was never short of nerve, and life with him was never ordinary. After a quick cook's tour, Bill started teaching Ben and me the ropes. I looked at tiny Justin and wondered how long it would be before he would be inducted as a crew member.

'Let's take her for a spin tomorrow morning', announced Bill.

I realized then that Bill had probably timed his purchase to coincide with the Easter break when his yachting friends headed out and anchored their boats in the spectacular bays and inlets around Pittwater, to drink and socialise.

'*Encore*'

The next morning, after dropping Polly off at a dog boarding kennel, I drove straight to the marina with the kids and enough clothes for a week, to meet up with Bill. An hour later, Bill shouted directions from the helm, while I released *Encore* from the mooring, and we were off on our first adventure. Bill was probably as terrified, sailing with a green crew on the yacht's maiden voyage, as Ben and I were, but I knew Bill was back in his element, with his sense of adventure fired up again. I realized then how devastating the drought had been for Bill, and how much he had ached for the thrill and freedom of sailing. His competitive nature had him itching to join his friends again to show off his new acquisition, complete with motley crew.

Bill took the helm firmly in hand and motored straight up Pittwater toward Lion Island until he worked up the nerve to hoist the sail. He seemed to be very attached to the sturdy diesel in the engine compartment below decks, since he counted on it to bail us out of any kind of trouble. While Bill hollered instructions from the helm, Ben and I hauled up and set the mainsail, and then the jib. Once *Encore* sliced through the water under sail, Bill cut the motor and we were on our way in the peace and quiet that all yachties cherish, with only the bow wave gurgling, 10 knots of breeze crooning through the rigging, and the sail gently crackling in the wind. Even Justin was quiet and enjoying the ride, safely below decks, in his basket.

Two hours later, we arrived at Refuge Bay and joined around 30 other boats that were already anchored in sheltered waters. Fortunately, I managed to pick up the mooring lines with the boathook after our third pass by.

With the boat safely moored for the night, Bill enjoyed his nightly rum while I fried up some delicious potatoes, Spam and onions in the galley. I didn't join Bill in a drink lately because, since the poisoning, even the smallest drink of wine gave me a sick headache.

Ben's maiden voyage on Encore

After dinner everyone, including Justin, felt mellow from the sea air and we all hit the sack for an early night. We slept well, except Bill, who was up on deck checking the spreader lights every hour during the night, and peering nervously overboard.

Sailing with a green crew was evidently playing on his mind and he was working out every imaginable plan B. Despite Bill's occasional fretfulness, we spent a relaxing week on board. Ben enjoyed rowing the dinghy ashore where he collected crabs and mussels among the rocks, and made friends with other boys from nearby boats. Some of Bill's friends recognized him on board and motored over in their dinghies to have a chat and a beer with him. The crowd consisted mainly of Sydney's captains of industry with a sprinkling of nouveau riche who had done well out of advertising, the professions and real estate. Even Bill finally relaxed when the trip turned into a boozy reunion with old sailing friends and business colleagues.

Eve enjoying Encore's afterdeck

One evening at sunset, after enjoying a thick fillet steak, Bill was nursing a rum and cola on deck, when I mentioned the bush in passing. I thought Bill hadn't heard a word, but the next morning he was up early and, over his morning coffee, announced it was time to go bush again. I half-regretted the decision, since I'd started to feel healthier and more relaxed in the fresh sea air.

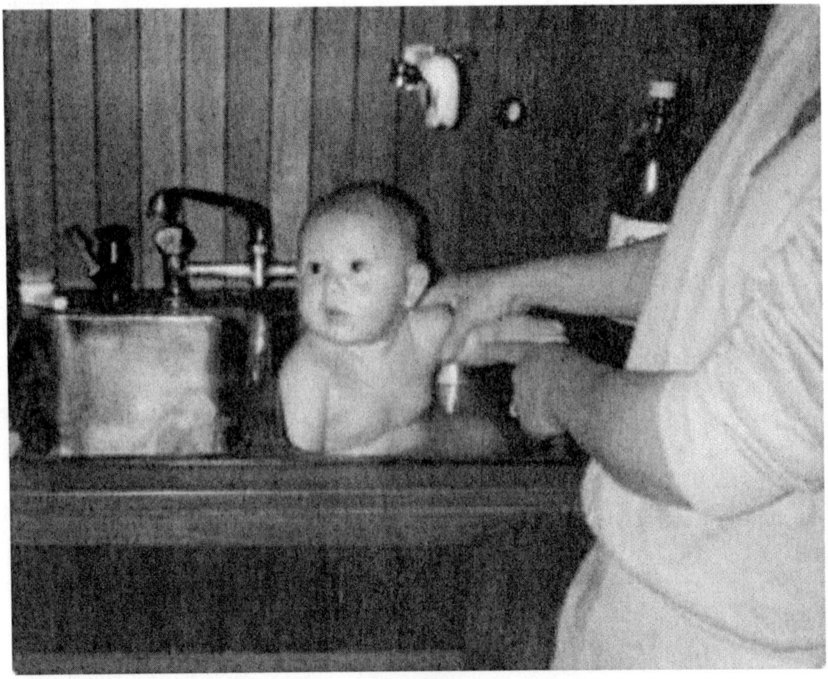

Eve bathing Justin in Encore's galley sink

After another long drive to Mudgee in two cars, the bush was an anti climax for us all. Bill preferred staying near the boat, I found Bayview less stressful and Ben wanted to stay around his best buddies, Peter and Jason. I was sure even Allen would have rather we'd stayed away, but now that Bill was back, Allen was on his best behaviour again. He turned up promptly and worked normally with young Darren, the farmhand, who had suddenly reappeared to report for work each day, now that Bill had returned. Darren was a tall, lanky boy with big hands and a healthy appetite for hard work. He said nothing bad about anyone, and trusted everyone. It wasn't hard to see that Darren's laid back nature could make him putty in Allen's hands.

When Ben returned to school after the Easter Break, I took Justin with me, in a sling, to help Bill. One morning we drove the Toyota Ute down to the stockyards where Allen and Darren were drenching the sheep they'd mustered out of the 200 acre paddock. Bill stood to watch the work, as usual. Judging by the small mob of sheep milling around the holding yards, it seemed the stock numbers were down, when we should have seen an increase after the rain. 'What's the tally?' asked Bill.

Since it was impractical to count animals while they were moving around the paddocks or holding yards, the stock manager usually counted them going through the drenching race, and recorded their numbers in the tally book. Allen slipped the book from his pocket and glanced at it intently, while impatiently wetting his lips. `They're all accounted for', he finally replied, first looking to Bill and then to me, as if to check our reactions. When we didn't respond, Allen continued drenching the sheep, while Bill and I took off in the Rover to check the boundary fences.

An hour later we reached the gate at the rear of the property, around 12 kilometers from the homestead. The gate was never used and led to a pot-holed back road to the abattoirs on the other side of town. The padlock that secured the gate had always been closed and was rusted from disuse. Only Bill carried a key for the lock and a spare went to the manager. Bill noticed something was different and got out of the Ute to investigate, only to find that the rusty padlock had been prised open, for the first time in years, and was now hanging loose between the two chain links that secured the gate around the post. Bill clicked the lock back together and gave it an extra tug to ensure it was securely clasped. We drove back to the homestead in silence. I said nothing, while Bill convinced himself that everything was under control.

In the next few weeks the property was running normally. All was quiet, and none of the goings-on that had happened while I was alone, had occurred while Bill was there. Before too long, Bill was off to Sydney again. He wanted to return to the boat - not to sail - but to do some varnishing and maintenance work on board. I couldn't blame him for wanting to pursue some pleasurable diversions after the stressful three years of drought. I longed to join him, but unfortunately I wasn't mobile enough to come with him; with a baby to look after and Ben attending the local school, I had to stay put.

After waving goodbye to Bill, I hoped that all would remain peaceful on the property, and I was able to enjoy the normality for the next few days, until Darren didn't turn up for work. By then I suspected that Allan told him to take a day or two off, to get him out of the way, and it was usually a sign that something peculiar was about to happen.

In May, the winter evenings grew dark around 5 pm and they were cold enough to start up the fireplace. Dinner was more fun to eat in front of a roaring fire and afterwards, Ben liked watching TV in his pyjamas until bedtime.

One night I'd finished feeding four month old Justin, and left him in the cozy winter room with his brother while I went to the kitchen to do the dishes. Before turning on the kitchen light, something outside the window caught my eye. I peered into the dark, down the private road, and saw lights flicking on and off near the stockyards. This time, I went outside to get a better look from my garden fence. Allen was not at his cottage and it was far too late, and too dark for him to be doing any work on the property, or down in the stock yards.

When the lights blinked again, I hurried inside and tried phoning Bill, but there was no answer at the Bayview house, and I thought he'd probably bunked in the boat for the night. I realised that the trespassers would be long gone before anyone could arrive to help me and that I would have to help myself until help came. I also remembered the constable's advice to get a license plate number as evidence. Just like last time this had happened, I was on my own again, but this time I refused to cower in my house like a victim; I was going out there to find out what was going on.

After checking on the kids, I rushed into my bedroom and felt around the top of the closet for the .22 and the box of ammo that I'd left there. My hands trembled as I wrenched back the bolt to check that the breach was empty. When I was sure the rifle was unloaded, I replaced the bolt and held the gun barrel up, while slipping half a dozen bullets, from the box of ammo, into my jeans pocket.

It was a dark night with just a sliver of a moon when I got into my car without a sound. With the car lights off, I drove slowly toward the stockyards and parked behind the shearing shed where I wouldn't be seen. From there I had to continue on foot, through a small grove of eucalypts, until I was in sight of the yards. I was relieved to shelter under a stand of gum trees, and took a few moments to compose myself before removing a bullet from my pocket and silently loading the rifle. With the barrel pointing down and my hand over the trigger guard, I stood for a moment to assess the strange and unfamiliar emotions going through me. I had loaded and shot various caliber rifles a thousand times before, while practicing target shooting and spotlighting for rabbits, and had never felt uneasy. But this was very different. It was the first time I'd had to handle a firearm to defend myself, my kids and property - and the seriousness of it suddenly hit me. I had no intention of firing a shot off at anyone, but I wasn't prepared to be a helpless victim in a dangerous situation either, nor did I intend being anyone else's target.

I took in a slow, deep breath and continued steadily through the grass, hoping that my steps would not be heard. Soon I was close enough to hear sheep bleating and men's voices talking in sharp bursts. I could make out the dark outline of the sturdy stock yard posts, but was still too far away to pick out any details. When the men's voices grew louder and more urgent, my heart raced and my hands were sweaty, even in the cold night, as I fought an impulse to run the other way. I had to force myself to move to the next clump of trees, closer to the stockyards, to get any chance of finding out what was going on.

From my closer position, a hundred meters from the yards, I spotted a truck backing up to the loading ramp. Its headlights were doused, but the rear brake lights flicked on and off as it maneuvered into position below the ramp. A flashlight beam illuminated the ramp for the driver of the flat bed truck. When the brake lights went off, the light beam moved and lit up the sheep race, which was full of stock waiting to be loaded. I realized these were the same blinking lights I had seen in the yards a few weeks before, from my kitchen window, and these were probably the same trespassers and thieves that I had chased down the road to our front gate. I stood in the dark for a few minutes watching the movement of sheep off the ramp and onto the truck, while several men called and yelled to each other. We had never loaded stock at night, let alone a Friday night. A serious crime was happening right in front of me, and apart from my word as a witness, I still had no hard evidence without a license plate number, and this was probably my only chance to get one. For that, I had to move closer to the row of trees closest to the yards. The mere thought of it made me shiver.

After taking a few moments to pluck up my courage, I placed my hand over the trigger guard, pointed the rifle safely upward and made for a cluster of three eucalypts growing nearest to the yards. I was now much closer than I'd wanted to get; close enough to smell the sheep in the yards. As I hastily took refuge behind the trees, my foot snapped a dry branch and made a piercing crack that echoed loudly. Hastily, I took cover behind the thickest tree, but it was too late: the men had already heard me and stopped what they were doing. Moments later a spotlight beam shone in my direction. I felt my heart beating wildly as I squeezed in behind the tree trunk.

'Who's there?' one of them called out.

I was now close enough to hear every word clearly but didn't dare move while the spotlight beam searched around the trees and lingered on the one I was behind. I stopped breathing in case my frosty breath on the cold night air gave me away. For a few terrifying moments I wondered what I would do if the men had seen me.

'Don't worry mate, we're done here anyway', said the other man in a broad ocker accent. Moments later, all was dark again and I was relieved to hear the tailgate being latched. Car doors slammed and the truck's diesel engine started up. I now felt game enough to peer out from my close position and was immediately rewarded with the evidence I'd come for. The numbers on the license plate were clearly illuminated as the men drove the truck, with our sheep on board, out of our stockyards. I watched the truck until its taillights were out of sight while reciting the six digits of its number plate.

Over and over, like a mantra, I repeated the number while driving back to the homestead, with my headlights off for good measure, just in case Allen was around to see me. It was the first bit of solid evidence that we had that something dodgy was going on.

Back home, the TV was still on in the winter room and I hastily wrote the numbers on a pad that Bill kept on his work desk, before I risked forgetting them. Fortunately, Ben didn't see me; he had taken himself to bed and Justin was fast asleep in his basket on the living room couch. I could unload the .22 and return unseen to my bedroom to put it safely away, back on top of the bedroom closet.

After checking on the kids, I tried phoning Bill again, but when there was no answer at the house, I barricaded the front door and went into a mind blank. My body had been in the trembling grip of adrenaline for an hour, and now that it was turned off, I was in shock and incapable of functioning beyond shuffling to bed like a robot, and passing out into a dreamless sleep.

Early next morning, a knock on the front door woke me up. Without a chance to recover from the terror of the previous night, I opened the door and was shocked to see Allen standing on my front door step. The sight of him made me instinctively wrap my dressing gown more securely around me.

He told me he was making a short trip to Sydney: 'For family business. Ya know what it's like'.

'When will you be back?' I asked.

'In about a week'.

'Where can you be contacted?'

'No place, really. I'll be stayin' with different relations. So I'll call Darren every night, just to see how things are goin' round 'ere', he said.

'What about the farm work?' I asked.

He avoided eye contact and stared out towards the driveway where his utility stood with its engine idling.

'That's all under control', he said impatiently. 'If youse need anythin', just call Darren', he added.

Anxious to get going, Allen tipped his hat like a country squire, jumped into his Ute and took off down the road. His sudden plans seemed odd, but I was relieved not to be alone with him on the property at night.

With Allen gone, I tried phoning Bill again, without success. Next, I called the police, which is what I had intended to do the previous night before going into shock.

On Saturday morning, the Mudgee Police were busy, as usual, processing the town drunks they had arrested the night before. The desk sergeant took my details and promised to send someone around, but it might not be until Monday, he explained, since the 'alleged' trespassers had already long gone and all the cruisers were out on other calls.

Despite feeling very unsupported and alone, I consoled myself with the thought that without Allen, I felt a lot safer. But I wasn't happy with Bill for leaving me to shoulder the lot; just when the kids and I needed him most, he was working on his boat and staying on board over night. In the days before mobile phones, that put him completely out of communication range. My only real consolation was the visit that the police had promised to pay me, and I hoped that would be soon. Meanwhile, I kept the kids close by me, and told Ben that I needed to know where he was at all times.

That night, after the kids were in bed, the phone rang, and I answered it with some attitude, thinking it was Bill. Instead, it was a gruff male voice.

'Who is this?' I asked, when the man didn't identify himself.

'Never mind who', he said. 'I'm just tellin' ya that your sheep are walkin' off your place on wheels – and you'd best keep an eye on your manager', he added, before abruptly hanging up.

I tried to think where I'd heard the man's voice before, but his gravelly voice could have belonged to any hard-wording, hard-drinking country man, who rolled his own. But I did take note of the man's warning, and knew that my intuition about Allen was fully justified. I was sure he had something to do with the men who were on our property last night, but the problem was proving a link between Allen and the thieves. Obviously, the man on the phone knew something, but chances were slim that an anonymous caller would come forward as a witness. Apart from the number plate of the truck on our property last night, which was different from Allan's Ute, the police would want hard evidence that Allen was involved, before they could do anything about him. The more I thought about it, the more I realized that Allen was as cunning as a sociopath, and it would take some mental gymnastics to be able to predict what he would do next. Figuring this out would take some time, I thought, and went to put the kettle on.

That night I sat down with a mug of strong tea and tried to think in the same convoluted way that I thought Allen would be thinking, if he really had something to hide. I figured this morning he'd acted like a guilty man who was making a two way bet. I thought he wouldn't have gone too far away; that he'd keep an eye on things from a safe distance by phoning Darren each day, and in case he was found out, he wouldn't ever be coming back. I was sure he wasn't acting alone and wondered if young Darren could be in on it. That seemed unlikely, since the boy had a heart of gold; it was more probable that Allen was using him as an unwitting pawn in his game.

It dawned on me that we had only one chance of catching Allen, and that was by discovering his exit plan, and we could only outwit him if we knew his game. Without enough evidence to stop him, he would become twice as cocky and dangerous with his drinking problem and his love of high powered weapons, and the property would be even more hellish for me and the kids to live on. Without Bill's help I felt I was left with no other option but to protect myself and the kids and figure out a way to finally stop Allen. Unfortunately, that meant I would have to gather my nerve and do the almost unthinkable.

Half an hour later I inched up the hill to the workman's cottage with a flashlight in my pocket. The cottage looked sinister as it stood in complete darkness, and the closer I came to it, the drier my throat got and the more riotous my imagination ran with thoughts that Allen might be in there waiting for me. Every thriller I'd ever watched came back to haunt me, but I took a deep breath and forced myself on toward the dark building. When the crescent moon emerged from behind a cloud and dimly lit the landscape, I took cover behind the spindly shrubs that grew on the hillside. From a safe distance, I watched the cottage for a while, and when there was no sign of life, I scurried up the path and crept up the front steps. Shivering with cold and nerves, I pressed my face against the dusty living-room window and squinted into the darkness. With nothing to see but pitch black, I prised the flashlight out of my overalls and shone the beam through the windowpane and into the darkness of the cabin's interior.

My heart nearly stopped when I saw a face staring at me — until I recognised my own reflection in the window. I steadied myself, and trained the beam around the living room, which turned out to be strangely empty of furniture, with only a few tatty old chairs and a nest of side tables strewn about, as if someone was trying to make it look lived in. Stale air wafted through the window's cracks, smelling of cigarettes and beer, like an outback pub. When I aimed the flashlight beam into the kitchen area, behind the living room, it revealed the chilling sight of an old lino-topped table with grease and dried blood trailing down its metal edge. I wondered if Allen had butchered the weekly lamb, to which he was entitled, on the kitchen table, or if I should be worried about a more menacing scenario. I felt a chill as I shone the light around the dining area which, like the rest of the cottage, contained no photos, knick-knacks or mementos of any kind; nothing to suggest that anyone was living there on a permanent basis.

Suddenly, from above my head came a loud crack. I looked up to see that the rusty eaves trough had disconnected itself from the fascia board and was now swinging back and forth in the breeze, with a possum clutching onto it for dear life, looking almost as terrified as I did. My first impulse was to take off down the hill to the homestead, but there was one more job to be done. This, I decided, could wait until morning, since my nerves couldn't take much more of this stress anyway.

I scurried down the hill and let myself into the homestead, locking what doors I could behind me. After checking on the kids, I poured myself a brandy, despite my recently discovered alcohol intolerance, and with shaking hands I took a shot in a single medicinal gulp. As the burning liquid trailed down to my stomach, I started to relax enough to connect the dots. It looked like Allen had already cleared out most of his belongings from the cottage, and would have no trouble leaving a few old sticks of furniture behind, but the true test of his plans lay in the whereabouts of his most cherished items, his chainsaw and high powered rifles. I figured he would never leave, permanently, without them. Tomorrow I would find out if he had taken them with him. If he'd left them behind, he would be coming back; if not, he would be able to stay away, if he needed to. His story about family business sounded like a convenient cover, and he was in an ideal position to monitor us, since he could phone us, but we couldn't call him. He would know exactly what was going on, since he kept in daily phone contact with Darren. I figured while he was laying low for a while, waiting to see if anyone was on to him, he'd left the cottage as if he were still living in it, to avoid suspicion. If he found out, from Darren, that we weren't happy with him, he'd leave his few paltry items in the cottage, and we'd never see him again. But I was also convinced that Allen had enough nerve to come back, if he thought we were none the wiser, and he could get away with it just once more. The more I thought about it, the more certain I was that Allen was laying a two-way bet on coming back. His chainsaw was usually locked up in the outside shed. I would take a look in the morning to see if it was there.

The next morning was Sunday and I nursed a headache from the single shot of Brandy I had taken the night before; the sheep dip poisoning had changed the way my body processed, even small amounts of alcohol, and just a few mouthfuls now gave me a major hangover. Unfortunately, though, there was no sleeping in for me. As soon as I had bathed and fed Justin and cooked Ben his breakfast, I looked to see if the coast was clear and headed back up the hill. Ben wanted to come with me, but I asked him to stay with Justin instead.

The workman's cottage seemed less menacing in the daylight, and fortunately it was still deserted. Behind the cottage was a small timber shed, and I opened the door to find none of the tools that were normally there; no chainsaw and no guns. Not even a garden rake.

It certainly looked like Allen was ready to bolt, and I figured our only hope would be to set a trap, lure him back and bring him to justice. To catch him we would need to act normally, as though suspecting nothing. Most importantly, we should not breathe a word to Darren so he would innocently continue to tell Allen, when he phoned each night, that things were as normal. I thought if the plan had any chance of working, Bill and I would have to do an award-winning acting job to pull it off.

Just as I wondered if my plan was even sane, let alone sensible, I was surprised to see a police car arrive at the homestead with a constable and a detective inside. It seemed they were particularly interested in the license number I had given the desk sergeant over the phone on Saturday morning, and came to get more details of the vehicle and the trespassers I'd seen on Friday night. After I had told them everything I had witnessed, the detective said that I had done very well to get the evidence, and that he would make some enquiries and get back in the next day or so.

After the police left, I finally got a call from Bill. He had briefly returned to the house for a fresh change of clothes, and to phone me, after spending the weekend varnishing the boat and sleeping on board over the two nights. When I started telling him what had happened, he was resistant at first.

'Don't be silly', he said. 'That's never happened to me'.

'Well, it's happened now', I said. 'Even the police were here'.

When Bill heard that, he packed up and headed for the bush the following morning. When he finally arrived on Monday at lunchtime, I was infuriated that he had left me in the bush alone to deal with a risky situation, and at the same time, I was glad and relieved to see him.

No sooner had Bill arrived, when the detective, who had been over the day before, drove up the road in an unmarked police car. After the detective talked to Bill for a while, the seriousness of the situation started to dawn on Bill, and he was outraged that someone was stealing our stock. He was also alarmed when the detective mentioned that he had looked up Allen's record, and after making a few inquiries, it turned out that the police were very interested in Allen. The detective told Bill he wanted to intercept Allen at all cost, and was keen to implement my original plan to set a trap.

For the next week, Bill and I pretended to go about our business as usual on the property. Darren turned up to work each morning. Bill gave him a list of jobs to do. At 5 pm Darren returned home. Each night, Allen rang Darren, just as he had said he would, and Darren innocently reported that Bill was back and everything was in order. Fortunately, Darren didn't even suspect our suspicions, which prevented Allen's finely tuned alarm bells from ringing. After a few days of nightly calls, Allen must have felt supremely confident that no one had suspected anything, and was brazen enough to send word through Darren that he would soon be 'back' from his 'family' trip.

Now it was my turn to rely on Bill to shoulder the responsibility that I had taken on my own. It was a long week, but when Allen finally turned up to work on Friday morning, we were ready for him.

Allen whistled as he checked in with Bill for the day's work schedule.

'Everything alright with the family?' enquired Bill casually.

'Yeah, no worries', replied Allen, looking relaxed, rested and cocksure of himself. The week with the 'family' had put an alcohol glow on his face, and he confidently headed out to the paddocks with Darren, in his Ute.

When Bill and I were sure Allen's Ute had disappeared over the hill, we phoned the police, as pre-arranged. Two senior detectives arrived a half an hour later in an unmarked car. We invited them inside, and I made them cups of coffee while they chatted with Bill and waited for Allan to return to his cottage for lunch.

Just before noon, Allen and Darren arrived back from the paddocks. I stayed safely inside the homestead and watched as the two detectives casually approached Allen, just as he got out of the Ute, and asked him to state his full name. Allen knew he'd been sprung as soon as he saw a couple of beefy, clean-cut men with short haircuts. I saw his jaw drop; he hadn't expected it. The detectives talked to him briefly before bending him over the hood of his Ute and clamping him in handcuffs. Darren, having no clue what was happening, watched horrified as the detectives led Allen to their unmarked car and drove him away to the Mudgee police station.

The police had dutifully carried out their obligation to inform Allen of the charge before arresting him, but we would find out, only later, what the true extent of Allen's crimes were, and why the police were so interested in him.

After Allen's arrest, Darren's parents came to see us, and begged Bill not to give Darren the sack. They insisted that Darren knew nothing about the scam, and it wasn't hard to believe. The young fellow was honest, but naïve and it was obvious that Allen had used him as a mole to get information. Bill accepted that Darren was entirely innocent and gave him a meteoric promotion: overnight the young man went from being a lowly farmhand to becoming a farm manager. In his new position he had the day-to-day running of the property on his young shoulders, and a number of casual farmhands working under his direction. Of course, that also meant that Bill and I would have to work along with them, and Bill's new toy would have to wait for him on the mooring.

End of an Era

All endings are also beginnings.

We just don't know it at the time.

— Mitch Albom

Bill liked having breakfast at his coffee table, with the calculator and morning paper handy. There, he had easy access to me in the kitchen, where he could read me snippets from the paper, and ask me to refill his coffee mug. One Saturday morning, he announced that we would be selling the property. I paused from drying the dishes and gave him a piercing stare through the doorway. I had heard correctly, he said, Landfall would go under the hammer. I suspected that he was tired of the never ending stress and demands of the property, and that the boat was calling him like a siren. I couldn't blame him, but wished that he would consult me before making life changing decisions.

As usual, Bill wasted no time, once he'd made up his mind. He listed the property with the local stock and station agent, and soon the news was all over the district through the bush telegraph.

The buzz attracted Stan, a cotton farmer from the west, who wanted to switch from cropping to grazing. Like Bill and I, Stan had battled through three drought years on his irrigated property at Narromine, where he grew cotton. The drought had lowered the underground water table and, for three years running, when he couldn't irrigate from his well he lost his entire crop, along with much of his topsoil.

Bill explained that in a good year, farmers can make big money from bumper crops, but all the running expenses and overheads come out of a single yearly cheque. In a bad year, farmers still have to buy seed and massive amounts of expensive chemical pesticides and herbicides to spray onto broad-acre crops. If the crop fails they're up for the overheads and still have to carry on planting for another year or two without a pay cheque. Two drought years could easily put a broad-acre farmer out of business.

What Bill didn't know, and I found out years later, was that organic farmers usually weathered droughts better than conventional farmers since they didn't have the added expense of buying toxic chemicals and their soil conservation techniques maintain moisture and minerals in the soil.

Stan however, was no organic farmer, and his plan B was to switch to grazing as a form of risk management. He figured that he could always buy and sell sheep and cattle in both good years and drought years, and have the income in regular intervals, instead of once a year. Bill and I didn't go out of our way to enlighten him on just how bad a drought can get for graziers too.

Stan arrived early one morning, with his wife and three grown sons, to look over our property. Bill took them on a short, half day tour in the Land Rover, since showing them around every paddock would have taken at least two days. At lunchtime they arrived back at the homestead, where I had cups of tea waiting, and soon they were having a lively conversation.

`Why would you sell a thousand acres under cotton?' asked Bill, trying to sound Stan out to see if he was a genuine buyer.

It was obvious Stan had once been naturally endowed with a large frame, but his body now folded like a wilted plant curling around the stem.

`I just can't do it any more', said Stan. 'My wife and I have arthritis, and we all have asthma and sinus problems'.

Dad reckons, in a drought, grazing isn't as hard as farming', said one of Stan's lads.

'With stock, you don't put all your eggs in one basket,' said the other son. 'Besides, Dad's back is nearly gone', he added.

Stan's wife agreed and added that she would like a tour around the homestead. For the next half hour, I showed them around every room in the house, except Justin's bedroom. It was one of those rare days when eight month old Justin went down for a nap, and nothing was going to get in the way of a few minutes of silence.

'I'll show you the room another day', I promised.

'That won't be necessary', said Stan looking at his wife. 'We like the place anyhow'.

With the family gone, Bill turned his attention to me. 'I reckon we've got a buyer', he said, with a self-satisfied grin.

'If they live long enough', I answered, before feeling guilty for the comment. 'Sorry, that didn't come out quite right'. I added.

Bill gave his high-pitched cackle that showed he was amused. 'Now you noticed that country people are always sick', he said.

I knew from my own experience that Bill was right, since I had never been sick in my life until I came to live in the bush. All kinds of birth defects and chronic illness raged through country families: cancer, asthma, arthritis and sinusitis. There was also depression, childhood behavior disorders and higher rates of suicide in country areas. Later, I discovered that as far back as the 1960s, Rachel Carson, the world-renowned genetic biologist, had documented links between prolonged exposures to farm chemicals and those disorders. Since then hundreds more researchers confirmed the link between these disorders and commonly used chemicals, but the message was slow to reach consumers who were never told know how to protect themselves from the chemical products that were causing their illnesses.

Later that afternoon we got a phone call from the detective at Mudgee police station. He told us that Allen had been officially charged with cattle and sheep stealing and his court hearing was listed for next month. The police investigation had turned up records that showed truckloads of our sheep and cattle had gone through the slaughterhouse, with the cheques, totaling over $60 000, written out to Allen.

I heard Bill ask the detective how he knew they were our sheep and cattle. The detective explained that the abattoirs kept ear-tags from every large stock consignment, and they were definitely identified as animals bearing our unique ear-tags. We were sure that Allen didn't count on the slaughterhouse keeping such meticulous records. He had even cashed the cheques, which gave the detectives plenty of evidence for a conviction. It was shocking enough to know that our stock was being stolen by the truckload, but there was more news. The detective added that Allen was running an organised stock stealing operation. His brother was a butcher in western Sydney and our lamb and beef had ended up on dinner tables all around Sydney's outer west. Bill was white with anger after hearing the news and was about to hang up when the detective added that Allen had a rap sheet a mile long, which included several previous convictions for sheep and cattle stealing. But worst of all, he had a dark history of violent crime, for which he had done time. The detective left it there and didn't elaborate. He would let us know the outcome of the hearing.

Bill hung up the phone and fixed himself a stiff rum. It was the first time he'd realised that Allen was capable of almost anything.

The news hit me even harder. My legs gave way and I slumped onto the couch. Now I was sure that it was Allen who'd shot our sheep and cattle; the ones I had found dead on the ridge. And I was also certain that the lights I had seen at night were Allan's Ute, as he drove around at night shooting anything that moved – for the perverse pleasure of it. He was a sociopath with a long criminal record and my gut feeling had been right about him all along. It was chilling to realize that I'd been alone for weeks, with a newborn baby and a ten year old boy, on our remote property, with a man who had a violent history, high powered rifles and a drinking problem. He could have done anything to us, and no one would have known about it for weeks. That terrifying thought made me nearly black out, until I noticed Bill staring at me.

'You're white as a sheet', he said.

'It's no wonder', I replied, as I took the glass from his hand and gulped deeply.

Bill looked at the empty glass and asked me to refill it for him.

'Get it yourself' I snapped 'I already do a man's job around here'.

Bill looked shocked. 'Do you think?' he retorted. 'What more do you expect from me?'

I leveled a stony look at him. 'I expect you to protect your wife and kids from predators', I answered.

With that sharp exchange, the balance between us had shifted, and we both knew it. I was no longer the naïve 29 year-old that Bill had dominated by the power of his personality.

Fending for myself had forced me to develop an independent mind of my own. That evening, Bill indulged in a rare display of generosity and took the kids and me to the RSL club in town for dinner.

We were both subdued the following week, as we contemplated how lucky we had been to survive our encounter with the violent criminal who lived among us. Fortunately, some good news interrupted the somber mood when Stan phoned to tell us that he wanted to purchase our property. He also wanted everything that came with it: walk-in, walk-out – except the horses. And, he aimed to move his family in by August.

Bill and I had mixed feelings about selling, since we both loved the property. He had owned it since the 1960s, while I had developed a strong, almost indigenous, bond with the land. But we both sensed the time had come for us to move on.

The next day, Bill phoned Stan back and agreed to the sale. That left us a month to pack up and move from Landfall. The challenge of selling the Arab horses fell to me. It was difficult to sell any kind of horse in the post-drought period, since feed was still expensive, and Arabians were a luxury that most people couldn't afford. Not even the big Arab studs were in a buying mood, since too many horses had glutted the market in the drought. But despite the depressed rural economy, I managed to sell every last horse - both the Anglo Arabs as well as the purebred Arabians that formed the breeding nucleus of the stud - though not for the prices Bill was looking for.

Right on schedule, in August, I loaded the last of the pure bred Arabians from the stockyards into two huge, commercial horse transports, and waved goodbye as they disappeared up the road. I felt silly waving to horses and certainly didn't expect them to wave back, but could think of nothing else to do while the tears welled up and my gut ached from the painful finality of seeing a part of my life, and years of hard work, disappear thought the front gate. I thought of Bill's late wife, Sally, who had bred the foundation mares, and was glad she was spared seeing the loves of her life leave Landfall, the only home they had ever known. Now I could only hope they would be happy and well loved by their new owners.

The next day, Bill and I loaded up our cars for the move to Bayview. We had taken only our personal belongings, and left the entire household behind for the new owners, since we needed nothing from the homestead that we didn't already have at Bayview.

Ben sat waiting in my Honda, looking thoughtful with the same mixed feelings that Bill and I were nursing. He had loved the life at Landfall: helping with the stock work, catching rabbits, searching for gold in the Fossicker's paddock and romping around the wool shed at shearing time. The bush had given him some of his most precious childhood experiences and memories, as well as irreplaceable skills and bush knowledge. In the land of Henry Lawson, Ben had come to know some colorful bush characters that inspired his own bush ballads about shearers, drover's dogs and the Mudgee races. On the other hand, the deprivations, drought and dangers in the bush had been hard on him too, and I could tell that a big part of him looked forward to returning to Bayview where he could spend carefree afternoons hanging out with his friends.

After securing Justin's basket across the back seat of my Honda wagon and placing him in it, I stood at our garden gate for the very last time, and looked back at the homestead. It seemed sad to see us go. So much of our lives had been lived there since Ben and I arrived, when Bill stood in the doorway, waiting for us to join him in our life together as a family. I remembered the countless trips with Ben to the school-bus, Bill's joy on the day the America's cup was won, bringing newborn Justin home to the homestead, only to see the first drought breaking rain, the sweet smell of the peppercorn trees in the summer, the wattle in the spring and the white snowfalls in mid winter. The Australian bush had a way of getting deep into the souls of those who lived there.

While Bill loaded Polly into his car, I tried to spot the goanna, but he hadn't made an appearance since that day, six months before, when he gave me a sign of the fateful days ahead. I said goodbye to him in my mind, before getting into my car and following Bill's Mercedes down the private road to the front gate, like a funeral procession. There, we stopped for a moment to look back, one last time, at the majestic and hauntingly beautiful Landfall. We both knew that it was the end of an era.

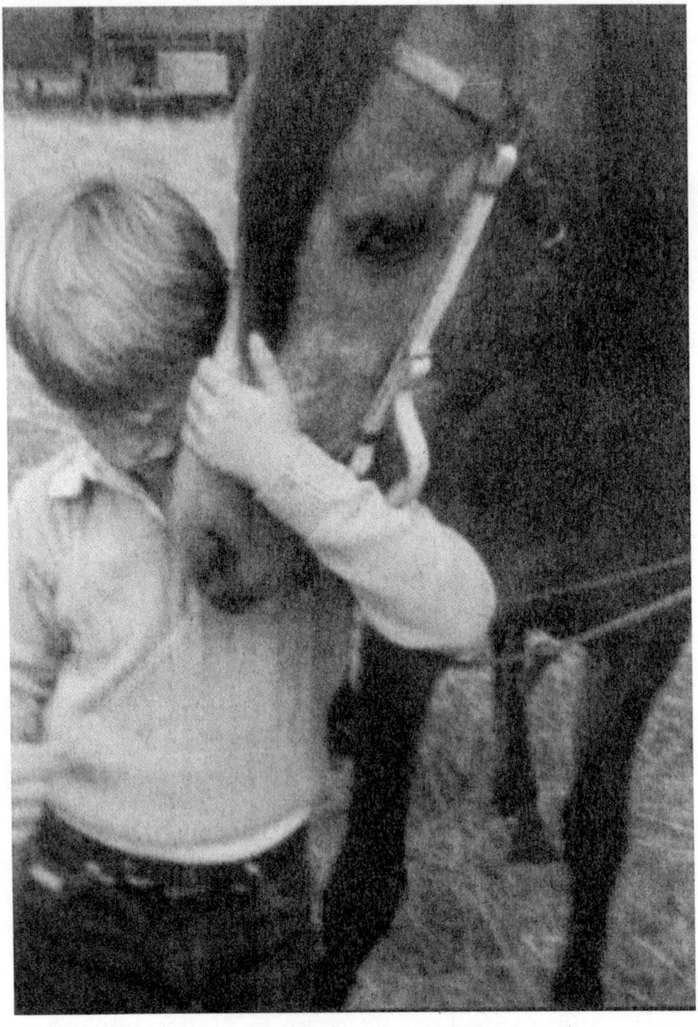

Ben says 'goodbye' to his Arabian colt

The Final Insult

O furious Wind, I am only a straw before you;

How could I know where I will be blown next?

- *Rumi*

Living out the rest of 1984, at leafy Bayview, was smooth sailing after the bush. I re-enrolled Ben in the local primary school at Mona Vale as a fifth grader, and each day after school he came home for a snack, before hanging out with his friends. Justin was nearing his first birthday and had settled down into a more regular sleeping routine that gave me a much needed few hours of rest at night.

Even Polly found a comfy mat on the balcony to sleep on all day. There, she spent a short, but pleasant, time that preceded the end of her life. She had never recovered from the organophosphate poisoning, and died that summer. Bill buried her in the rocky slope of our backyard. The kids missed her terribly and, to fill the gap in their hearts, I brought home a pet kitten for them; a cheeky black and white male, that Bill named Ratsey, after the famous sail makers, Ratsey and Lapthorne.

Despite the easy lifestyle, I still had traumatic memories and flashbacks of being alone in the bush, of Allen's night time stalking, and having to wield a rifle, to protect my kids from trespassers, stock rustlers and criminals. Unfortunately, Bill had little patience to hear me out.

'Forget it. It's over now', he would say, when I brought up the subject. His remedy for everything was to go out to sea, and for a while, his therapy even worked on me. I gradually relaxed after we spent every weekend and most holidays on *Encore*, anchored in the idyllic coves and inlets of Pittwater. We often brought friends, relatives and even neighbours aboard, and Ben invited his school friends on the sailing trips. When Justin grew too big to have his morning bath in the galley sink, I took him to baby swim lessons, and soon he was at home splashing around with me and Ben in the shoal waters of Yeomans Bay, when we anchored there.

More than ever, Bill enjoyed the company of his colleagues and sailing mates who motored over to us in their dinghies, and spent the long and balmy summer evenings on board *Encore*, reminiscing about their ocean racing victories in the Sydney to Hobart and Admiral's Cup ocean races. We entertained yacht club commodores, captains of industry and the ocean racing glitterati of the day.

For me, it was a time of desperately needed rest and recuperation. Even though Justin was still a handful, I knew the sea air was doing him good, and even I was starting to recover some of my energy with only a few bouts of asthma to occasionally slow me down.

Eve with Justin and family friend Chris on board Encore

Just when it seemed that we were all starting to recover from four years in a toxic agricultural chemical environment, we acquired a flea problem in the Bayview house. Instead of discussing it with me, Bill acted on his usual knee jerk impulse and called in the pest controller. A few days later, I came home with Justin from the supermarket, to find a pest-control truck parked in our steep driveway. The driver had wedged two house bricks behind the rear wheels to stop the truck from sliding down the drive, with its plastic tank full of pesticide on board. I stood on the patio, with a hungry baby in my arms, and watched through the French doors as the pest controller sprayed inside the house, with Bill supervising and making sure he saturated every object with a pungent smelling chemical.

Since the poisoning, I was determined to find out which chemical was being used around me and the children, and what antidote, if any, it had. When I called out to Bill from the patio to ask him about the chemical, he came out to wave me away with an annoyed expression that told me he would have preferred the job to have been done before I came home, to avoid this kind of questioning from me. Instead, I was determined to get an answer, and went over Bill's head by motioning the pest controller directly to come out. Looking puzzled, the pest controller skulked on to the balcony with an unlit cigarette dangling from his mouth, clutching his plastic pesticide tank. I used the opportunity to ask him what he was using to fumigate our house.

The pest controller switched from looking puzzled to sheepish. 'We're not really meant to say, but it's quite safe', he insisted as he demonstrated its knock down power by aiming the spray nozzle at his own socks, and spraying the chemical onto a cluster of fleas that had landed there from the carpet.

'See? Knocks em dead', he beamed, when the fleas staggered at first, and then dropped dead in their tracks. 'And it's armless to yoomans', he added, with satisfaction.

'It really stinks' I complained, as the sweetish chemical odor made me back away towards the edge of the balcony. Justin started whimpering, before working up to a scream. 'Do we have to keep the windows open after this? I asked.

'I can't smell it anymore, cause I'm used to it', shouted the operator over Justin's wailing, 'but I spose it's a good idea'.

'What is that stuff anyway?' I asked.

The pest controller hesitated, and primed the pump again to re-pressurise the tank. 'Alright lady, not that it'll mean much to youse — but it's Dichlorvos', he said impatiently.

Just then, Bill stepped in and impatiently waved me away with a quick sweep of his arm. 'Go on, get out of here for a while!' he hollered across the balcony, before gesturing for the operator to continue the job.

Bill was right about that part. I needed to get Justin away from the house during the fumigation. 'You stay out of there too!' I answered back to Bill as I returned to my car and strapped Justin, kicking and screaming, into the rear car-seat.

I idled the next few hours away with the baby, in the Baker's Marina car-park, watching the boats sail by on the foreshore, until he had calmed down. At 3.45 pm I intercepted Ben at the bus-stop on the main road to ensure he didn't wander home into a cloud of chemicals. To pass even more time, I took the boys to Elaine's pie shop at Mona Vale for some home made meat and vegetable pies, which we ate on a park bench behind the public library. Afterward, Ben amused his baby brother on the swings, while I sat and watched from the park bench, and thought about what to do next.

Four hours later, around 6 pm, I took the children home, only to find that the operator had left no follow-up instructions or safety information with Bill. Today, pest-control operators using toxic chemicals, are required to make sure the premises are ventilated for at least four hours after the fumigation. With most other chemicals, we would have been relatively safe to come back, as we did, after four hours. However, unbeknown to us, Dichlorvos, an organo-phosphate pesticide, is so toxic that the re-entry time is not four hours, but four days. Neither the government regulator, nor the pest control company informed consumers of this fact, and in the days before the internet we had no way of doing our own research, and were sitting targets for being chemically poisoned for the second time in a year[2]. Without the necessary information, we resumed our usual family routine four hours after the application, on carpets still damp with Dichlorvos. I prepared dinner in a kitchen with surfaces covered in the toxic chemical, and we breathed the chemical vapors in the air. That night, I had the most serious bout of asthma yet, and would have ended up in hospital, had the Ventolin puffer not worked just in time to allow me to breathe again.

---

[2] In Australia, Dichlorvos was withdrawn for domestic use in homes and surface sprays in 2007 and is approved for use only in grain silos

After the fumigation was all but forgotten, its damaging health effects were just beginning, especially since we had been primed by the toxic exposures to agricultural chemicals the year before. A few days after the fumigation, Justin became extremely irritable and developed a fever, while Ben came down with a virulent case of tonsillitis and enlarged lymph nodes[3].

While trying to nurse both boys, I also came down with the worst illness I could remember, and collapsed into bed with a fever (typical of OP exposure). For the next week I couldn't stand up straight without fainting, and had to crawl to the bathroom on all fours, and inch my way back to bed. My head was throbbing with one long migraine headache and I felt nauseated, disorientated and delirious. Since I was too ill to function, Bill took care of the children for a while, but soon he came down with a bout of diarrhea and developed sciatic pain in both legs, making it hard for him to soldier on.

Once I was out of bed again, I made doctors' appointments for all of us, since our recovery seemed too slow and the cause of our mystery illness was still undiagnosed. Our GPs were part of a family practice consisting of a father and son GP who both had different approaches to medicine. The elder was authoritarian and didn't welcome questions from patients, while the younger, who was often booked out, had a more inclusive approach that made the patient feel more like a per-

---

[3] Hypothermia and Fever After Organophosphorus Poisoning in Humans – A Prospective Case Series Journal of Medical Toxicology. 2010 Dec: 6(4)379-385

son. After several appointments with both GPs, neither had answers about what was causing our health problems, nor could they explain why everyone in our house had become suddenly ill, after our house was sprayed with pesticides. Instead, both doctors treated our symptoms; Ben was prescribed antibiotics each time his glands or tonsils were swollen, Bill got pain killers for sciatica and anti-diarrhoea medications for the runs. I was prescribed various migraine preparations for the bad bouts that came almost every week, but there was nothing the doctors could do about the fatigue that drained my energy and left me weak and lethargic.

A year previously, we had rarely visited the doctor, and now at least one of us visited the surgery almost every week. It was the constant tiredness that sent me back to the doctor, time and again, since it interfered with my life the most. After the tenth trip to the doctor, he wrote me out a prescription for an anti-depressant. I headed across the road to the chemist to have it filled, and as I waited in line at the counter, I was gripped by a strong urge to leave without it. Hadn't all our health problems started at once and weren't they physical? I tore up the prescription and left the pharmacy, determined to find the real cause of the problem, and have it treated, if it took me the rest of my life.

When I came home that afternoon, Bill asked me what the doctor had said. I wished that I'd had some answers for him and for me.

'Maybe you should have taken those tablets', he said, when I told him what I had done with the prescription for anti-depressants.

'I would have, if I thought they could fix my headaches and asthma, or stop me feeling like somebody twice my age'. I answered.

That evening, we had Bill's long time friends around for dinner. I'd spent hours preparing baked Atlantic salmon, and along with it I served a fine Chablis. Everyone enjoyed the dinner, except me. After only half a glass of wine, I felt suddenly ill and practically blacked out at the table. I vaguely recall Bill helping me upstairs to bed. The next morning I was violently ill with a nasty migraine. Apart from vomiting and a splitting pain in my head, I felt embarrassed and puzzled. I'd never had a physical problem with alcohol before the sheep dip poisoning and had always been able to drink without getting a hangover. I had abstained from alcohol during pregnancy, but a few months after the first poisoning in the bush, after Justin was born, I had noticed my alcohol tolerance gradually reducing until now, when I couldn't even have a sip of wine without a violent reaction. I had to face the fact that for some reason, my body could not process alcohol as before and that the mystery illness in my body was forcing me to abstain from alcohol completely. This proved especially trying, at first, during social occasions, or when we entertained our friends on the boat.

After making a huge effort to radically adjust my life-style, I expected my health to improve, but after a few months of tee-totalling, my migraines, asthma and fatigue only got worse, and I wasn't the only one getting sicker.

Several months after the fumigation, Justin became so distressed one morning that I thought he must have been in severe pain, when he didn't give his usual irritable cry, but a continuous bellow. He flailed his arms and legs about, and his tiny, pale face was covered in a lather of sweat. When he refused to eat or drink that day, I took him to the local hospital, where we waited in ER for several hours until a doctor was available to examine him. By then, Justin's tummy was rigid and he looked dehydrated. After an abdominal x-ray revealed a dilated bowel with distended pockets of air, the doctor decided he needed intra venous fluids and searched for a vein to insert a tiny butterfly cannula. Before he could put up a drip, the doctor was called away to another child that had arrived in the next cubicle, as an emergency.

While comforting Justin, I caught a glimpse of the other baby, about Justin's age, through the curtain. His little body was white and motionless while the doctor tried to insert an airway and a nurse thumped his tiny chest, in a last ditch effort to resuscitate him. But the child made no sound, and in a few minutes the staff stopped their frantic attempts at resuscitation, and all was horribly quiet. Somewhere was a mother beside herself with worry and grief. I picked whimpering Justin up and held him close to me, and gave thanks that he was alive. After a few more hours waiting, a doctor transferred Justin to the children's ward for overnight observation. His small arms were black and blue with bruises from unsuccessful attempts to find a vein, but he still had no IV after refusing all food and fluids for the past 24 hours.

After spending a sleepless night at home I returned to the hospital next morning only to find Justin lying in his hospital cot looking the textbook picture of dehydration, with dry doughy skin and pale sunken cheeks. When there was still no diagnosis to explain his abdominal pain, I signed Justin out of the hospital and took him home, where I made up a half liter of electrolyte solution and persuaded him to drink it in small amounts over the day. He was still listless for a day or so, but to my relief, he soon bounced back to his cranky self. That incident proved to be the first of many bouts of abdominal pain of unknown origin. After several more such episodes, I was somewhat relieved when the younger family doctor diagnosed abdominal migraines; at least it was a diagnosis, but unfortunately it came with no treatment regimen, since no one knew what caused it and what to do about it.

Meanwhile, Justin started displaying the beginnings of the severe hyperactivity and impulse control problems that he would suffer from during much of his childhood. Again and again he would take a running leap and try to fling himself over the balcony rail into the jasmine bushes below, or repeatedly attempt to launch himself overboard from the boat. He even jumped out of the dinghy when we rowed ashore from the mooring. Fortunately, I'd purchased a harness and tether for him, and we clipped his safety harness onto the mast when we were on the boat. I had no prior experience with this behaviour, and had never run across it with Ben or the children I'd encountered while working in paediatrics during the 1970's. The best I could do was apply the normal motherly discipline, show him the boundaries and do whatever was necessary to keep him safe. In frustration, Bill would ask me: 'Why can't you control him?'

I had no answers and was desperately trying to understand what was happening to us. Over the next few years we beat a well worn path to the family doctor, who said, about Justin's hyperactivity, that 20 per cent of children are 'little devils', and he must be one of them. Unfortunately, the 'little devil' diagnosis didn't explain his rashes or frequent, painful bellyaches.

About me, the doctor said that it was normal to get sick and tired; my headaches 'won't kill' me, and to keep taking either aspirin or paracetamol for them. I had no option but to follow this advice if I wanted to be functional enough to handle the daily chaos that our lives had become.

I began to notice more frustration in Bill as well, which he vented on me and on Ben. On Ben's eleventh birthday, I wanted to host a little party and buy him something special. It meant I had to ask Bill for extra money and that never went down well. Despite our outward affluence, and Bill's good income from investments, he had kept me on a meager household budget, and the kids and I often had to make do with sausages or mince, while Bill insisted on steak. I had tried every tack, but was unable to persuade him to increase the budget to a reasonable amount that allowed for clothes for me and the kids, and a few extras from time to time. In desperation, I dipped into the child endowment money the government gave mothers. That $20 per week gave us no end of relief to the constant penny pinching I had to do to get by each week.

On Ben's eleventh birthday I had used child endowment money to buy Ben a skateboard, which I wrapped up as a surprise for the next day. The next morning, Bill started an argument with Ben about the skateboard, which escalated into such a scene that I took the kids and sought refuge with a family down the road, who put on a little party to help us celebrate Ben's Birthday with them. When we returned, Bill had calmed down, and later that night I let him know that more displays of aggression toward Ben would not bode well for our marriage. From then on, I was careful to protect Ben as much as I could from the fallout of Bill's frustration.

Our lives seemed to be collapsing before our very eyes and there were simply no answers to the difficult issues that cropped up every day.

When Justin was three years old, I decided to take him to a paediatrician, once again, in hopes of getting some answers to his health problems. I made an appointment with the family GP for the evening surgery session, where I intended to ask him for a referral to a baby doctor, for Justin.

I'd had another headache that day, and took a couple of aspirin before going to the surgery. In the waiting room, Justin was his usual hyperactive self. While I checked in with the receptionist, Justin made a dive for the corner table and threw a stack of magazines on the floor, before heading for a crate full of toys. One by one, I had to stop him throwing stuffed toys around the surgery. One set of furry dice narrowly missed an elderly woman and bounced off the wall. As I mumbled apologies and tried to divert Justin's attention, I felt a searing hot rash creeping over my scalp and neck, and it spread like wildfire down my chest, to the rest of my body. It was intensely itchy, and my attention was momentarily on scratching. While trying to restrain Justin, I noticed the usual judgmental glances. It wasn't hard to guess what people were thinking. One man moved away from us and sat in the corner with a look that seemed to say, 'You don't need a doctor, lady, you need a cage for your kid!' I could tell that people were attributing Justin's behaviour to bad mothering, instead of a disorder affecting the child.

Just as we were the centre of attention in the room, I broke out in a clammy sweat and was suddenly in need of a toilet - for both ends. I felt so nauseated and faint that I left Justin in the waiting room and struggled alone down the hallway. I barely made it to the ladies room, where a glance in the mirror revealed a mass of ugly hives covering my face, neck and body. Between the raised wheals, my skin was deathly pale and oozing beads of cold sweat. I went into the stall and knelt to be sick into the toilet bowl. After the last heave, I felt faint and buckled to the floor. I don't remember how long I lay on the cold tiles, drifting in and out of consciousness, but after coming around, I crawled back to the waiting-room on all fours, since each time I tried to stand up, I felt faint and collapsed again.

I felt every eye on me when I crawled through the door of the waiting-room in a daze, and looked for Justin. The receptionist must have alerted the older GP, who peered out from his consulting room and motioned me in, and onto his exam table. The man watched from a distance, as I struggled onto the exam table, while he slipped on a pair of latex gloves. By then, my throat felt thick and I found it hard to swallow. I heard a loud ringing in my ears, and the doctor's gloved hand was the last blurry image I saw before passing out cold.

I didn't know it yet, but I had just experienced my first anaphylaxis: a life threatening allergic reaction characterized by hives, shock, unconsciousness, which, if left untreated, often ends in death. To me, anaphylaxis felt like going under an anaesthetic, and despite heralding the onset of life threatening circulatory collapse and anaphylactic coma, the feeling of blacking out certainly offered some relief to the nausea, burning and throat closing sensation.

The next thing I remember was regaining consciousness after the doctor had administered an intravenous injection into my arm. He looked relieved to see my eyes open and focusing again, and told me that he'd given me intra venous hydrocortisone and an antihistamine injection for the anaphylactic reaction. He assumed that I knew about my severe allergies, but I told him that, although I knew about anaphylaxis from working in the ER, I'd never gone into anaphylactic shock in my life, before today.

'Well, you'd better find out what's caused it then', he said, almost abruptly. 'And stay away from it like the plague'.

'How can I – without killing myself trying?' I mumbled, while cautiously raising my head to see if I would pass out again. My question stayed unanswered, since the GP was already on the phone to Bill. 'You'd better pick your wife up from the surgery straight away', he said. 'She's not very well'.

I was grateful to the doctor for saving my life, since without emergency treatment for anaphylactic shock I would have died. But I found out later that I should have been admitted to the hospital for tests and observation. In anaphylactic shock the blood pressure drops so low that there is risk of circulatory collapse, brain damage and kidney damage. (Smith1982). Patients need medical follow-up after anaphylaxis and important instructions on what to do in case it happens again. It would have been helpful if the doctor had tried to find out why a previously healthy woman suddenly developed life-threatening allergies, but no such follow-up was done, and I was not referred to a specialist allergist. Unfortunately, the GP was busy with his evening list and wanted to go on to his next patient without spending more time on me, and I was about to walk out of his surgery, and into a minefield of potential anaphylactic triggers, without the tools to save my own life. In those days, there was virtually no help for people who had been poisoned and suffered from the kind of chemically induced immune system disorders that I was developing.

The doctor showed me to another room where the receptionist had amused Justin after I'd collapsed. I was happy to see him again, but with all the drama, his problem still hadn't been attended to, and would have to wait for another day.

When Bill came to collect us from the surgery, I had a thumping headache and needed Bill's assistance to walk upright without fainting. It was a relief to get home and climb into bed. Bill looked after the kids, since I was unable to do anything, apart from lie in bed in a daze.

That night I woke up every hour, wondering if I would lapse into shock again, and reflecting on what could have caused the anaphylaxis, and what would have become of my kids had I died or lapsed into a prolonged coma. It was especially worrying, since I hadn't a clue what trigger had caused the anaphylaxis in the first place.

In the moonlight I looked at my husband beside me. He looked peacefully asleep, but I knew we both felt bewildered, overwhelmed and frustrated at the chaos surrounding our lives lately. I loved him deeply, but something was tearing our lives apart, and for the first time in my life, I felt that I was not in control of my destiny, that I was being forced along a path that was not of my choosing. I knew that most normal relationship issues had resolutions, but these strange health problems had the power to tear our lives apart, and they seemed to have no solutions. If bad things didn't stop happening, I wondered where we would end up, and if we could weather the storm.

Life in The Farce Lane

Pain and foolishness lead to great bliss and complete knowledge -
For Eternal Wisdom created nothing under the sun in vain.

- Khalil Gibran

Bill and I tried to deal with life's challenge, each in our different ways. For him, only going to sea made difficult emotions bearable. When the Sydney to Vanuatu Ocean race came up, Bill jumped at the chance to gather up his old racing crew and take off on the 2537 km race. On race day morning, he left with his sailing bag and waterproofs and was, as usual, full of enthusiasm to meet the weather and the sea head on. To him, nature was always more predictable than the vagaries of human relationships. After the confusion of the past few years, I could see his point, but while some of the unfolding issues that threatened our relationship were beyond our control, others, such as the money issue, were fixable. Unfortunately, Bill refused to compromise, and once again left me short of funds while he was away at sea.

As usual, while Bill was gone, I stayed at home to look after the kids. Ben was now in his first year of high school, where he'd expanded his peer group, from Jason and Peter, to include 20 other friends with whom he went on to form life long friendships. He had grown into a sensitive and intelligent young man with a shock of sandy hair that made him look like a surfer. I had nurtured his musical abilities, and after buying him his first guitar, I drove him, each Saturday, to guitar lessons. Before too long he formed a band, played lead guitar and sang vocals. After a few school performances, he and the other boys in the band had no shortage of female admirers for the rest of their time in high school. Ben developed his musical ability to a high level and could easily have become a performing musician but, after leaving high school, he went on to study for a profession.

Justin was now old enough to go to nursery school and I enrolled him in a preschool, at nearby Narrabeen, for two days a week. From the beginning, he found learning difficult, due to his undiagnosed and untreated chemical brain injury, which continued to cause him problems with hyperactivity and impulse control. Despite this, he seemed to benefit from socialisation and the added stimulation of toys and learning materials

With Justin at preschool, I had a few hours of free time each week and was able to reconnect with some friends and revisit a few hobbies that I'd put aside. Occasionally I went out with the girls and, after taking drawing classes, I started painting portraits and even restored an old oil painting of Bill as a child. During this time I first noticed that I developed headaches and asthma when working with turpentine and oil paints.

I was getting other unpleasant reactions around foods too: Coffee made me ill and gave me a racing heartbeat, bread bloated my stomach, foods with artificial additives, and especially Chinese food, triggered a migraine.

Not long after Bill went to sea, I was at home alone, working on a painting, when I stopped to make myself a ham sandwich for lunch. Half an hour later I felt itchy hives erupting on my neck, and spreading rapidly over my body, accompanied by the cold sweats, and nausea. It was the terrifying, but now familiar, sensation of lapsing into unconsciousness just before an anaphylaxis. Despite realizing that this was a medical emergency, I was incapable of driving myself to the hospital. As before, my blood pressure dropped so low that any attempt to stand upright ended in blacking out. I discovered that lying down flat on the hallway carpet allowed me to stay conscious for long enough to phone an ambulance. After that, I passed out.

The next thing I recall was waking up on a trolley in the ER of the local hospital. The doctor had just finished giving me a large dose of adrenaline, and I was still shaking from the injection.

'What do you think you're allergic to?' he asked.

'I don't know', I said, still stupefied from the anaphylaxis. 'I just ate a ham sandwich and blacked out'.

The doctor was surprised when he heard that after my first anaphylaxis the GP had not referred me to a specialist allergist. He told me I was playing Russian roulette with my life if I didn't have access to the proper emergency medications to deal with another anaphylaxis. He insisted I go to a specialist allergist the very next day, and he even made an appointment for me before discharging me from the ER. Since I had no access to funds while Bill was gone, I hoped the specialist would accept Medicare.

That night I was still shaky from my trip to the ER, when after dinner, the kids and I heard, on the TV, that tropical cyclone Manu was forming north west of Vanuatu, and was expected to cross the path of the yachts on the Sydney Vanuatu Ocean Race. Since Bill was leading the pack in *Encore*, he and his crew would be the first to be hit by the 80 km per hour winds and high seas. Bill and I were out of communication, but every other day I was able to find out Bill's latitude and longitude from the Cruising Yacht Club (CYC), and could then plot his progress on my chart of the South Pacific. According to these calculations Bill's last location would intersect with Manu in 24 hours, if he did not steer away from the island archipelago of Vanuatu. Changing course, away from the target, was counter-intuitive to Bill, who had a killer instinct to win ocean races at almost any cost, but on the other hand, punching a yacht through tumultuous high seas in cyclonic winds meant risking the lives of all those onboard. That night, I prayed that Bill would receive the best available weather information and that he and the crew would be guided to safety.

The next day, when I got the kids ready for school, Justin must have picked up on my worry. 'When is Daddy coming home from Van Tomato?' he asked. I found Justin's concern for

Bill and his child's pronunciation so sweet that it served as a tension breaker while I taught him how to say Van-u-a-tu.

It was set to be a stressful day until Bill's location at sea was released from the CYC late that afternoon. Meanwhile, I dropped Ben off at his high school and drove Justin to the Narrabeen preschool before going on to North Sydney for my appointment with the allergist.

As soon as I walked through the allergist's door, and told him of my two anaphylactic reactions, he took my condition very seriously and carefully took down my medical history. Together we deducted that I had my first anaphylaxis after taking aspirin for a headache and that yesterday's emergency was probably triggered by a reaction to ham, or more specifically, since I had no problem with pork, it was most likely to a chemical preservative such as sodium nitrate, used to preserve the ham. The specialist was the first health professional to acknowledge that I had become highly allergic to certain chemicals, and to understand how scary it was for me to manage this potentially fatal condition on my own without the help of emergency medications. He prescribed injectable adrenalin [4] and also an adrenalin inhaler for me to keep close to me at all times, in case I had another anaphylaxis to a chemical that I did not yet know I was sensitive to, and he advised me to avoid, 'like the plague', the chemicals that I knew I was sensitive to. By

[4] Referred to as epinephrine in North America.

then, I had experienced death's door, and was living on the edge each day.

I wanted so much for life to be predictable again as it was before. 'Is there a cure for this?' I asked hopefully.

'There is no cure for anaphylaxis. You have to face the fact that you have serious life threatening allergies. I can't say for sure what caused it, but I can say for sure that you need to either avoid the triggers or manage your allergies - or the anaphylaxis will eventually kill you', said the allergist. 'I can desensitize you to many substances such as fur or pollen, but not to chemicals', he added.

The man was kind and concerned, yet direct and professional. Despite being still shocked and horrified about my diagnosis, I trusted his no-nonsense approach, and was relieved to finally get help. At least some of my questions were thoroughly answered, and he treated me like an intelligent person, but most of all, I was hugely relieved to know that in future I had emergency medications on hand, and would not be at such a high risk of dying during my next anaphylaxis. When the consultation with the allergist was over, I went straight to the pharmacy where I spent the remaining household money to have the two prescriptions filled: one was for an EpiPen (adrenalin auto-injector) and the other was a Medi-Epihaler (inhalable adrenalin puffer).

When I got home that afternoon, I phoned the yacht club and asked for *Encore*'s latitude and longitude, which I plotted on my map to get Bill's latest position. From the co-ordinates it was obvious that he had turned around 180 degrees and was now sailing in a south westerly direction. He had clearly taken steps to avoid the more northerly path of cyclone Manu, as it headed across the Pacific, toward the far north coast of Australia. Despite the defensive maneuver, I was under no illusions about the difficult conditions Bill and the crew must have been experiencing on board the yacht. Sailing *Encore* so close to a cyclone would have been like being in a front-loader washing machine, and I wasn't giving up on prayers for their safety, just yet, until they were back on dry land.

When Bill walked into the house a week later, with a salt encrusted two week beard, I was torn between being angry with him for leaving me short of funds yet again, and being grateful that he was alive.

For a while, I forgot about being angry, until we were invited to the yacht club for a dinner in honour of Bill's contribution to ocean racing. I asked him for extra money, since I had no clothes that were suitable for the formal occasion, and I needed to buy an evening dress. When this fell on deaf ears, I resigned myself to not going. I was tired of the arguments about money, but I had reached my limits, and decided that it would be the last time this would happen.

On the night of the formal dinner, Bill put on his dinner jacket and black tie, and asked me why I wasn't getting dressed. When I told him why, even he realized how his attitude toward money was affecting our lives, but by then the stores were closed and it was too late for me to get a suitable outfit.

Bill went to the dinner on his own, which turned out a magnificent gathering of champion ocean-racing skippers, crewmembers and their wives; while I stayed home with the boys. That night, Bill came home from the event sheepish and apologetic, an attitude he rarely took with me. I was in no mood to be forgiving, but determined to change the situation permanently.

'It won't happen again', promised Bill.

'No, it wont', I agreed. 'Tomorrow morning I'm looking for part time work'.

Bill looked horrified. 'I'm not having any wife of mine work!' he thundered. 'What would you do anyway?'

'Back to hospital work', I answered, 'to a career that I've been highly qualified in, since long before I met you'.

The next day, despite Bill's protests about not wanting me to work outside the home, I left him in charge of Justin and drove into the city to apply for a position at the Royal North Shore hospital. There was a shortage of intensive care trained RNs and I walked out of the hospital administration office - hired for part time duty on the sixth floor ICU suites. On my way home, I wondered if my health would stand up to the stressful job, but I was more confident now that I had my emergency medications, and I thought I could manage two days per week.

I arrived home to find Bill in a panic, and bloody tea towels were strewn around the living room. Bill had not often looked after Justin by himself, and was unprepared for the extent of his hyperactivity. When Bill had taken him down to the marina, Justin jumped into a concreted tidal pool at low tide, when there was no water and he hurt his chin and shattered several teeth. When I saw that Justin had a gaping wound under his chin, I took him to a surgeon that I knew, that afternoon, for stitches.

I spent the next year struggling to balance work, kids and the household, and trying to deal with my impaired health. I worked my hospital shifts mostly on Justin's two preschool days, but sometimes Bill would have to look after Justin, and he didn't always enjoy the restrictions this responsibility imposed. Ben, on the other hand, now spent much of his time hanging out with his friends.

On my days off, I increasingly reconnected with old friends and took Justin with me on visits. In warmer weather, Joan, an elderly girl friend, often invited me to her home, near the water, at Palm Beach. One particular weekend she hosted a small get together and invited us, so I took Justin for a swim to drain him of excess energy, before joining the party at Joan's house.

Not long after we arrived, Joan put out a dip with corn chips. I was usually very careful to avoid problem foods when I prepared meals at home. When I was out, I always asked about the ingredients before eating food that anyone else had prepared. In the past year, I'd encountered dozens of unexpected anaphylaxis triggers, and had to self-administer the emergency adrenalin medication after developing anaphylactic reactions to a long list of foods that contained nitrites as preservative. This included: ham, bacon, sausages, a variety of small goods, soup mix, stock cubes and gravy mix - as well as many other fast foods and processed foods that contained the chemical, but did not specify it on the packaging. In addition, I'd had serious reactions to a variety of commonly used medications such as aspirin, paracetamol and codeine. I had done well to avoid dying from allergic shock by using the adrenalin on dozens of occasions during the first year, and I thought I had a handle on the triggers and could avoid them from now on. But on this particular day, I assumed the dip was made only of cream cheese, and got stuck into it. Within a few minutes I felt the first itchy sign of hives, and to my horror, realized that I'd left the adrenalin at home. Now I had to make a split second decision, and figured it would take too long to explain to anyone what I needed; my house was just a few minutes away, while an ambulance would take half an hour to arrive, and even longer to take me to the ER.

I left the party without explanation, raced home in the car with Justin, and struggled up the driveway and into the house, just as I was starting to see black and beginning to hear the ringing in my ears, before passing out. I'd left the adrenalin in my other handbag, which still stood on the hallway glass table. While lying on the entranceway floor, I primed the injector and sank the needle, through my clothing, into my thigh.

I came around from the anaphylaxis to see four and a half year old Justin staring at me from the doorway. Shakily, I cleared away the needle and syringe, and reassured him that all was well. I regretted driving home in that condition, and putting my son in danger, but it was a hard call to make.

Later when I'd recovered, I phoned Joan to explain the reason for my abrupt departure, and to ask her what was in the dip. It turned out that she had made it with cream cheese and added French onion soup mix that contained sodium nitrite as a preservative.

When I couldn't enjoy food and drink with others, it became increasingly obvious that my health problems were socially isolating, and this was becoming more apparent when we were on the boat.

By that time, our only time together as a family, was on board *Encore*. Sometimes I took my sketch pad and tried to capture the elusive lights and colours of Pittwater, but company was never far away. Ben invited his friends on board and Bill enjoyed the company of his yachting friends as well. Most of his crowd had their own vessels, and those who didn't, Bill invited on ours. Naturally, I was the chief cook and bottle washer to an ever expanding group of Bill's friends, but some of our most relaxing times were when Bill and I took the kids out sailing on our own. In the afternoons, when Justin went down for a sleep, and Bill retired to the navigator's bunk for a nap after his rum and cola, I would have the helm to myself. Ben and I liked cranking up *Encore* and tacking up Pittwater in a stiff breeze. My idea of a good time was to give Ben the helm when he could enjoy sailing, without Bill yelling directions at him. Instead, most of the time, I had to wear a permanent smile, chatter with guests for half the night and serve the hors-d'oeuvres, since Bill always lit up when his friends came on board.

One long weekend Bill invited Charlie and Kay out to Yeoman's Bay with us, since they were without a boat of their own. Bill and I were really quite fond of them. Although Charlie was a fearless sailor who had crewed on some well known international ocean racing yachts, his business life was less illustrious. Charlie had made and lost several fortunes, and at that time he was in a downward spiral, with the banks closing in. For Charlie there was only one cure for his troubles: go to sea, have a few drinks and retell his extensive repertoire of sailing stories. Poor Kay had heard them all before, and after devoting forty years of her married life to keeping Charlie in line, she gave up on the thankless task, and started giving Charlie a run for his money in the story and drinks department.

We set sail with Charlie at the helm, who grasped the wheel, spread his legs the full width of the cockpit, and sniffed the air like an old schooner captain about to round Cape Horn. Bill was setting the sails, while Kay sipped champagne. Although I hadn't been able to enjoy a single drink for a few years now, my job was to fill the guest's glasses, and keep the food coming, while keeping Justin safe and Ben amused.

A few hours later we picked up a mooring at Yeoman's bay and our arrival soon attracted friends from the boats moored nearby. Somehow, I managed to cook dinner for an extra ten guests who boarded not long after we arrived. Unfortunately, my food allergies made it impossible for me to eat much of what I'd prepared for the guests, and I had no time to prepare a separate meal for myself. Our guests, however, enjoyed themselves so much that they stayed into the small hours, before disappearing in their Zodiac dinghies singing naughty sea shanties. Bill and I surrendered our double cabin to Charlie and Kay, and slept in the saloon, while the boys slept in the forward double cabin.

The next day Charlie and Kay emerged from the rear cabin, a little hung-over, but sniffing the sizzling bacon and eggs I was cooking in the galley. Before I had a chance to finish grilling the tomatoes and serve up breakfast, Charlie had downed half a dozen cans of Foster's lager.

'Heart starter', he muttered, when he caught me looking astonished. He went on to finish the case after breakfast as a bacon-and-eggs chaser. At 10 am Charlie decided it was time to open a couple of bottles of white wine as a pre-lunch aperitif.

Later, three bottles of sturdy Hunter Valley red accompanied lunch, while last night's bleary-eyed crowd boarded once again. I felt like a short order chef and babysitter as I kept the food coming and Justin amused. Ben, meanwhile, took the dinghy ashore and explored the beach for the afternoon, while Charlie and Kay kept up the pace all day – from Chablis to Chardonnay, from Hermitage to Cabernet Sauvignon, topped off with flagons of rough red that someone on a nearby boat had ferreted out of their grog locker.

Finally, when the sun set over the yard-arm, Charlie brought out his personal stash of over-proof rum and the long yarns of lunchtime turned into one-liners that had the crowd rolling with laughter.

Though Bill always drank moderately, he too enjoyed the fun and wasn't inclined to set limits on his friends. However, at two in the morning, after I'd announced that the kids and I needed to go to bed, Charlie and Kay launched the dinghy, and cast off with the remaining diehard crowd to do the rounds of the other yachts and plunder their rum lockers, since ours had run dry.

Bill and I crawled into our bunks by two-thirty and slept for a couple of hours until Charlie and Kay woke me up at dawn with a rendition of 'What Shall We Do with a Drunken Sailor?' as they rowed the dinghy back to our boat. Next, I heard a couple of splashes as they'd lost their footing on the boarding ladder and took a dip into the bay. Just when I wondered if I should get up and fish our guests out of the water, I heard the sound of our marine toilet being pumped, which told me they had boarded safely in their wet clothes, and hopefully were getting ready to hit the sack for some sleep.

Marine toilets have been the downfall of many ships, and I wondered if they'd remembered to close the sea cock that pumps water in, from the bay, during the flushing process.

From my bunk in the saloon, I heard Kay giggling in the rear cabin. 'Oh look, darling, the head's all sparkling', she said, referring to the seawater phosphorescing in the dark as it refilled the toilet bowl.

'Don't be so bloody stupid, woman, and come to bed!' came Charlie's reply.

Bill was still sleeping soundly and I must have drifted off to sleep again myself, only to have some bad dreams about the boat sinking. A while later I awoke with a jolt and bolted out of my bunk when I noticed the odd way the yacht was moving; from her usual light, bobbing sway around the mooring, she now felt like a sack of potatoes. I got up and peered through the galley window to see a piece of seaweed floating by, not too far down from the window. This could only mean that we had taken on water. 'Get up', I yelled at Bill. 'We're sinking!'

He immediately woke up from a dead sleep and slid open the engine compartment, only to discover seawater lapping at the bottom of the engine casing. The bilges were full, and the water level was nearing the cabin-sole.

'Get everybody up and pumping!' he shouted, after brusquely throwing two metal pump-handles toward me. I had to wake Charlie and Kay, turn off the water intake valve of their cabin toilet which they'd left open, and start pumping the bilges by hand, while Bill started up the electric bilge pump to empty out the water in the boat. By now the boys were awake and it was bedlam, as we took turns pumping out the bilges to prevent the boat sinking.

After the last drop was back in the bay where it belonged, the party was over, two days early, and Bill released the mooring line for the trip home. While Bill motored down Pittwater, Charlie and Kay sat in the cockpit looking ashen and seedy. Back at our marina mooring, our guests packed their bags and said their goodbyes.

'We had a marvelous time Bill', said Kay, with a dazed expression.

'Absolutely fantastic', mumbled Charlie, his eyes still glazed.

We helped them into the dinghy, where Ben was at the helm, waiting to ferry them across to the marina. Halfway across someone must have shifted their weight which capsized the dinghy. Bill and I stood watching helplessly from the deck twenty meters away, and were hugely relieved when Ben's head surfaced, followed by the others. Kay's handbag came up for air as well, and she retrieved its contents from the water while Charlie tried replacing his comb-over back onto his head, after he'd emerged from the water with it coiled around his neck.

The hero of the day was Ben, who had righted the dinghy by himself, helped everyone back on board, and delivered our guests safely to the marina.

When Ben finally came back on board soaking wet, I wrapped him in towels, and gave him a big hug. I was incredibly proud of my 16 year old son for being cool headed in an emergency.

Unfortunately, Bill didn't join in the praise. Instead, he vented on Ben whenever he felt frustrated or confused. That day, it spoiled being on the boat together; the only thing that still united us as a family.

Stormy Weather

God and I both knew what it meant once;
Now God alone knows.

— Friedrich Klopstock

I enjoyed my shifts at the hospital. For two days a week I could do a job for 8 hours, and walk away from it knowing that I had helped someone. But at home, our issues seemed to be a permanent fixture that neither Bill nor I knew how to tackle.

One day Bill announced that we were returning to the land. He planned to purchase 80 acres at Newry Island, near Coffs Harbour, where he wanted to breed cattle and commute the six hour journey back to Bayview, on weekends and holidays. I knew it was Bill's way of trying to recreate the happiness we'd had at Mudgee in the early days of our marriage.

This idyllic property was surrounded by the fast flowing Kalang river, and came with a lovely homestead surrounded by mango trees and banana palms, but the sacrifices that each one of us would have to make proved to be the plan's eventual undoing.

The day after our arrival I enrolled Ben at Bellingen High school where he was unhappy from the beginning, without his peer group at Bayview. Meanwhile, I had to leave my work at the hospital, give up my financial independence and rely, once again, on Bill for the household money. This triggered our previous issues, and for the next year Bill increasingly took his frustrations out on me.

Meanwhile, Justin's hyperactivity escalated out of control. He was in his last year of preschool before starting primary school the following February, and was having problems learning, socializing and sitting still, even for short periods. I realised that his poor impulse control was liable to get him into serious trouble one day. This was apparent when I took him shopping at the local fruit market one day, where he wriggled around in the trolley while I bought some fruit and vegetables. As usual, I hurried to the checkout before he played up too much. A police officer, also shopping there, was waiting in line behind us, and while I was paying at the counter a male voice roared behind me. I turned to see the policeman adjusting his duty belt around his waist.

'Is this your kid?' he demanded.

I nodded. Surely Justin couldn't be in trouble with the police already, I thought.

'He went for my holster', he said. 'Sorry I shouted', he added in a quieter tone.

Briefly, Justin was stunned into silence before impulsively reaching for it again. This time the officer stepped back, and I admonished Justin, but my correction made things worse. He reflexively started working up to a tantrum, and there was no stopping him once he'd started.

We'd been down this road many times before when people stared, raised their eyebrows or shook their heads. Sometimes they laughed or grimaced, others made disparaging remarks, but no one actually offered any assistance that would help me get through a queue faster. I felt like a fugitive; constantly on the run, skulking out of shops and public places.

This time, I made a dash for the car and tried to fasten Justin into his seatbelt, and throw the shopping bags into the back seat, so I could make a fast exit from the car park. I noticed that a woman had pursued me out of the shop, and caught up with me just as I was about to back out of the parking space. When she paused at my window to catch her breath, I dreaded what she was about to say.

'My dear', she said, 'I wanted to tell you that I had a boy just like yours, and he grew up to be a wonderful young man. Whatever you do, just don't give up'.

I was so overwhelmed by this random act of kindness that I sat in the car and cried for the next half hour. It was a massive release of built up stress. By then, Justin had finished his own performance and was watching me intently. When I'd finished crying, I made the decision to do whatever it takes, and pay whatever price necessary, to find a way of helping Justin. And I also promised to provide Ben with a more supportive home, now that he was growing up into a young man and trying to find his own place in the world.

That night, I asked Bill to go with me to a marriage counselor. When he flatly refused, I took the kids, left the Newry Island property and headed for a friend's house in Sydney. She was a naturopath and her herbal remedies helped calm us down during the few weeks we spent with her. Meanwhile, I returned to my part time job at the hospital, and after some peace talks with Bill, he agreed to go to counseling with me, while we tried reconciliation.

We agreed to relocate the family back to Bayview, where Bill and I went to a few sessions with an excellent therapist. While I poured my heart out, Bill could not understand the issues involved. He was a man's man of the old school, who thought he was providing very well for his family, and couldn't understand the need to give emotional support.

We might have reconciled our differences, had it not been for the challenges of my unfolding illness and the special needs child that Justin was; a fact that Bill was still in denial about. He couldn't understand why our lives had suddenly gone off the rails, and frankly neither could I until years later when I realised that we'd been facing challenges that were well out of the ordinary, without the necessary information to handle them. To make matters worse, Bill still remained unwilling to compromise on even the run of the mill issues such as household money, and without his co-operation I knew that we had come to a dead end in the marriage.

After only a few months of reconciliation, Bill left on a solo holiday to New Zealand, and I took only a few personal belongings from our Bayview home, and left the marriage with the children.

From now on I was on my own, and the next years proved some of the most difficult of my life. Bill was resentful that I had taken the decision to leave, and he would show his anger by being combative in the courts. For the next three years, until the settlement, I had to survive on my own, without access to funds, and support myself and the kids on my hospital salary.

When it was over, the kids and I emerged from the fight for our survival as traumatized as any refugee from a war.

Baptism of Fire

Change is not made without inconvenience,

Even from worse to better.

— *Samuel Johnson*

The final divorce settlement was enough to put a down-payment on a home, while the other half went on legal fees, but at least we were now able to move forward. Ben was in his final year of high school and Justin was in the third grade, and my first impulse was to provide the kids with the kind of home that we'd had in Bayview, and continue giving them all the benefits of middle-class stability.

In 1992, Australia was in the grip of the recession that Treasurer Paul Keating said the country 'had to have', and times were tough. But the economic downturn offered me a fantastic opportunity: when the real estate bubble had burst I had the chance to buy a run down house, in an affluent area, for a bargain basement price. I found a two storey, three bedroom home; a renovator's dream that stood on a generous corner block and offered a filtered view of Pittwater. But in order to afford the mortgage payments and cover the household expenses and the kid's needs I would need to step up my hospital workload. Since the kids loved the house and wanted it, I told them it would only be possible to achieve if we all worked together over the next year. This meant that when I worked early shifts, Ben would need to help out by taking Justin to the school bus-stop with him in the morning, and on my two afternoon shifts, he would need to be at home with Justin.

After I'd bought the house, all went well for a few months, but since Ben was in his final year of high school he preferred to hang out with his friends. He had bonded closely with his peer-group, and stayed with their families, often for several weeks at a time, instead of coming home.

It was a difficult time for all of us, Ben was trying to grow up, we were all shattered after the past few years, and I was desperately trying to keep what was left of my family together. Without seeing much of Ben that year, I felt our family was breaking up completely. I blamed myself for not being able to give my kids more of a sense of family cohesion because all of our relatives lived overseas, and since I had to work full time.

For the next two years I struggled on with baby sitters for Justin, as well as before-and-after-school childcare, which cost half of my salary. The heavy workload gave me little spare time to keep track of Justin's school progress, but occasionally I had a chance to stop by the school playground, where I found him wandering around on his own, looking isolated and lonely. As soon as I could, I enrolled him in the local soccer team, hoping the sport would drain some of his hyperactive energy and connect him with other boys. Through this, he made a few lifelong friends.

Justin's true talents however, lay with animals. Unlike people, they sensed the kindness of his heart and responded to his affection, while he in turn responded to their non judgmental devotion. Soon a ring tailed possum moved in with us, and a flock of Sulphur-crested Cockatoos lined up on the balcony railing each day waiting for Justin to feed them biscuits and sunflower seeds. The birds wanted Justin's exclusive attention and ganged up on our black and white cat, warning him not to mess with them. The cat must have taken the lesson to heart and never hunted birds in his lifetime; only rodents and the occasional baby brown snake, which he left, still wriggling, on the door mat for me to inspect. For some reason he didn't want to finish them off until I'd had a chance to see his catch. I never knew what I would find when I woke up in the morning before getting ready for my shift. Animals always had a soothing effect on Justin, unlike school where he seemed to have a harder time, than other kids, achieving the work that was expected of him.

At work, in the intensive care unit, I often received phone calls from the school principal who wanted me to come to school immediately and discipline Justin for his latest misbehaviour. I was unable to leave work, since I usually had an unconscious patient on life support to look after, but the man still liked to call me about every thing. One day, while I was managing a dying patient, the principal phoned ICU to let me know Justin had broken wind in class. There wasn't much I could do about it at the time, but a few years later I learned the real reason for Justin's abdominal migraines; they'd started shortly after the second chemical exposure, which had damaged his gut lining. This resulted in food intolerances and impaired digestion. In short, he had intestinal inflammation and his food fermented rather than digested in his gut and created bloating, which he could hardly do anything about, while he sat in class.

Fortunately, for the previous three years Justin had teachers who had devised a reward system that would allow him to earn gold stars for acceptable behaviour. It was easier for him to aim for an incentive than it was for him to learn from the constant punishment and detentions doled out by those who didn't know that he had very little voluntary control over some of his impulsive behaviours. This year's teacher however, did not use that reward system with Justin. Instead, she went to the principal whenever Justin showed impulsiveness or hyperactivity. When I finally had a conference with the principal, he gave me an ultimatum; either medicate Justin or find another school. The school counselor gave me the number of a local paediatrician and strongly advised me to contact him.

While I was thinking over my options, I came across an article in a woman's magazine, written by a paediatrician, with a checklist of signs and symptoms of attention deficit and hyperactive disorders (ADD and ADHD). Justin had every listed symptom of ADHD, and I checked off each box. A few days later, I made an appointment with the paediatrician recommended to me by the school counselor.

The doctor took only a minute to decide that Justin had every aspect of the disorder including: inattention, lack of control, impulsiveness and hyperactivity. After a few more tests, the paediatrician applied for the necessary ADHD medication, and we waited two weeks for the health department to approve it.

For Justin, this was the first move forward since he was born. It was a huge relief, after watching my son suffer for the past nine years, to find out that this insidious and isolating disorder had a name, even though no doctor had confirmed the cause of it. Justin had been robbed of a normal life for his entire young life, but even though he now knew he had a disorder, he was a long way from accepting it.

'Mum', he would say, on a bad day, 'I don't want to be like this'. Justin was nine years old when he first voiced the distressing insight that he was living in a world that was out of focus and after hearing that, I was willing to try anything at the time, including medication, to get him some relief.

The first tablet had a dramatic effect on Justin. Within fifteen minutes, he turned pale. Just when I thought he was going to be ill, he picked up a book and read it from cover to cover; something he'd never been able to do before, even for five minutes. It was a large print book about the adventures of an earthworm, and after closing it, Justin walked around the house as though seeing it for the first time. Later we sat and talked sensibly, when I we could never have a complete conversation before. This time he was focused, but very pale.

For a while, Justin improved, until the side effects of the drug kicked in. Over the next few months he became allergic to his pet hamster and I had to re-home it. He also stated getting migraines after taking the medication, and it became obvious that one day we would have to try a different approach. But at least he glimpsed that his life could change, and that he had taken a first step in a journey that could lead him toward his true potential.

By the end of 1993, Ben had finished his high school final exams. He was unsure about his future plans and, while deciding which university he wanted to attend, he got a job at the local servo and hung out with his band and his peer group, practically full time.

I still had a strong urge to provide my boys with the best home and lifestyle possible, and was determined to renovate the house, even without help from anyone, to make that family life real for them.

I worked up the courage to apply for a bank loan, and when it was approved, I started the renovations on my own. For the next few months I came home from work, put on a pair of overalls, and chiseled tiles off the bathroom walls, ripped up the carpet and painted the walls. I hired a tiler to lay new bathroom and floor tiles as well as Italian terra cotta over the concrete on the patio. I installed a new kitchen and had a new carpet lain throughout. To top it off, I designed a garden and had it landscaped with Australian native plants. The house now looked cozy and welcoming, inside and out, but now I had another bank loan to repay and that meant even more shifts at the hospital. Even though the house looked fantastic, I still didn't see Ben more often than before.

Instead of enjoying the renovated house, I began getting severe bouts of fatigue that landed me in bed on my days off. I didn't realize that my home renovation effort had exposed my system - already weakened from the two serious pesticide exposures - to glues, solvents, paints and a cocktail of toxic chemicals, and they were now taking their toll. After a stressful day at work, my limbs felt like lead and I wondered if I would even make it home without pulling over and resting at the side of the road. My allergies spread to include household cleaning products, and I started getting rashes from most cosmetics and personal care products. My migraines were now so frequent and severe that I had to take days off work, and I was beginning to wonder where all these health problems were leading me, and whether I could even stay financially afloat or physically able. Doctors continued to be of little help, except to manage some of my symptoms, but none of them could tell me what was causing my illness.

Mid January 1994 was hot and dry. A scorching hot westerly had wilted Sydney for the past week. I dropped Justin into holiday child care at 6.30 am, and rushed to the hospital for an early shift. ICU was climate controlled and I was determined to enjoy the air conditioning before sweltering in the car again, on my way home. After handover, the unit manager assigned me to my patient; a 17 year old girl with a rare and incurable malformation that affected the vascular structure of her brain.

Alice's pretty, heart shaped face peered out from an enormous head bandage. She was just a few days post op after having a burst cerebral aneurism surgically clipped, and it was in such an inaccessible area of her brain that the neurosurgeon hadn't expected her to survive the surgery. She was far from being out of the woods though, since the surgery had fixed only one of many weakened blood vessels she had bulging in her brain. In fact, she had so many vascular brain abnormalities, that it was impossible to remove them all, and she was living with the risk that a weakened vessel could burst at any time and end her life within a few fatal seconds.

Alice had the kind of courage that inspired people. She was aware of her diagnosis, and of her bleak and uncertain prognosis. With a virtual time-bomb ticking in her head, she was determined to live her life to the fullest, no matter how much time she had left.

When I wasn't checking Alice's IVs and the monitors above her bed for vital signs, and plotting them on the flowchart, I sat beside her. When she stretched out her hand, I held it. She told me of her ambition to be a journalist, and of her doctor's veiled warnings that she might not realise those aspirations. I thought that if anyone could achieve the impossible, it would be Alice. But for her, life was lived only one day at a time.

Just as I began to realize how lucky I was despite all my challenges, Alice asked me if I had a secret ambition that I'd never told anyone about. When I said I did, she looked curious. For one so sick, Alice displayed no self-absorption, and I felt I could open up to her question. I confessed that most of all I wanted to help my son to recover from his illness; I wanted both my sons to find happiness, and I wanted a close knit family.

Alice looked at me with soulful eyes and smiled.

'And some day I want to write a book'. I added, now feeling embarrassed that I'd let my guard down and blurted out my private stuff to a patient.

'That sounds really nice', she said. Her voice was still hoarse after the being on the ventilator. 'Do you know how to make those things happen?' she asked.

By now, I was intrigued. 'How?' I asked.

She leaned toward me to whisper the answer. 'If you want it badly enough, imagine it, pray for it, and it will happen'.

'How can you be so sure?' I asked.

'Because that's what I did', she replied. 'That's why I'm still here'.

In the days before personal development gurus, this girl gave me the perfect formula to manifest change in my life; get clear on the goal, imagine it, pray for it, and take it forward one day at a time. It made instant sense, and I was overwhelmed that this sage girl had suddenly appeared in my life to deliver this perfect wisdom. I turned away and reached for the box of tissues on the windowsill. For a while, I stared out of the sixth floor window until I'd composed myself enough to get on with my job of helping Alice for the rest of my shift.

While driving home across the Roseville Bridge, I found myself stuck in peak hour traffic, and the heat wave had practically melted the asphalt, but I was deep in thought about what

Alice had said. My forty second birthday was in a few weeks time, and my life needed a major overhaul. I took Alice's advice and prayed for changes that would bring me closer to what I hoped for, and left it up to the Universe to decide what those changes might be.

Whoever said: 'you don't always get what you want, but you always get what you need', could have been referring to what happened next in my life. As I neared my home on the Northern Beaches, an eerie orange glow bathed the landscape from the sun that hung in the sky like a huge copper sphere. The smoky haze of a bushfire was heavy in the air by the time I collected Justin from the child care centre, but I was too tired to worry about it. Probably the fire department doing some back burning or putting out a spot fire, I thought. Meanwhile, there was dinner to prepare, and washing to do, and another shift in the morning.

I was happy to see Ben home for dinner that night. Over a plate of spaghetti, he told me that he was enrolling at university for the fall semester, but it was in the north of the state and he would be living on campus. It was great news that my son had decided to study for a profession; I was very proud of him and, of course, Justin and I would miss him.

After dinner Ben picked up his motor bike helmet and took off down the road to his friend's house, while I forced myself to overcome my tiredness and tackle the washing up. It was now dark outside and for a while, Justin and I watched a mindless sitcom. Just as I was half asleep on the couch, a friend phoned from South Australia.

'Are you guys all right?' she asked, sounding worried.

'Of course we are', I answered. 'Why do you ask?'

'We've been watching the bushfire on TV, and it is unbelievable!' she said excitedly.

'Yeah, I know – there's one every year. Probably a bit of back-burning', I said wearily.

When my friend seemed unconvinced, I glanced out of my study window and watched, in total disbelief, as a tall stand of gum trees on the hill, across the road, exploded into naked flames and spewed glowing cinders high into the night sky.

'Bloody hell, I've got to go', I shouted. 'It's coming this way!'

I slammed down the receiver and racked my brain about what to do next, before deciding to look for some news on the TV. Just when I'd changed the channel in search of a news bulletin, a power failure plunged us into darkness. I'd depended on the media to tell me what to do, but now I had to handle the situation by instinct and common sense.

Justin latched onto the waistband of my jeans while I felt my way around in the dark to find some candles and matches, since we had no flashlight. At that stage, he was more afraid of the dark than the fire, but we were both spooked when the cat let out a mournful cry and followed his sixth sense to a place in the southern corner of the house. Animals can discern reliable information from the environment and I should have headed there myself. Instead, I wandered around the house by candlelight, worried about Ben's safety and gathered up my most cherished belongings, including photo albums and old letters, and placed them in the downstairs hallway ready for evacuation.

While deciding whether to evacuate, I tried the phone and the lights again, but the line was dead and the power was still off. When blue-grey smoke poured through the open windows, I went through the house to close them. After we could do nothing more inside, Justin and I stood on the front porch to see what other residents were doing. Most had joined the convoy of cars, laden with people, pets, and surfboards, and were heading out of the neighbourhood. The hillside was now well ablaze, with volatile eucalyptus trees exploding in a shower of sparks, and flames leaping high into the sky; but despite the terrifying scene, a few residents remained behind to save their homes. This gave me renewed hope and enough courage to go outside and hose down the garden to prevent my own trees from catching alight. When the outside was thoroughly soaked, I decided to take Justin with me on a fact finding mission. With more information, I could make better decisions and since I desperately needed to buy a flashlight, I figured I would try to make it down to the supermarket at Mona Vale.

Not far from our home, we encountered the first roadblock. I asked the female constable where it was safe for us to go, but she didn't answer my question and waved me on with her flashlight instead.

Instead of going straight ahead, I pulled over to the curb to watch the heavy traffic of evacuating residents converging from the surrounding areas. It was obvious that, once I'd passed through the roadblock, the officers weren't going to allow me back home, and I still had Ben unaccounted for. I figured, either the police didn't have much information, or they weren't giving it out to the public. With police and fire sirens screaming in the night, the disturbing atmosphere of chaos and panic filled me with more fear than the prospect of facing a fire, so I did a hasty U-turn and chose to go back home and take my chances. There was no getting my hands on a flashlight, but as a consolation, I now had the car radio that broadcast news about the fire.

Unfortunately for me, the reports focused on the fires in the south of Sydney, when it was also blazing through my area on the North Shore, including Ingleside, West Head and Ku-ring-gai Chase, which was well alight, all the way up to Gosford. I urgently needed to know about the fires close to me before deciding whether to evacuate, since there was no point leaving one place and going toward another fire.

On the way back to my own neighbourhood I noticed only a few stalwart residents out with their water hoses. On impulse, I continued along the Bayview and Church Point foreshore where a close knit community of yachties lived. The residents there were putting out spot fires in their yards with garden hoses and buckets. Further down McCarrs Creek Rd, Justin and I joined a group of locals. Together, we watched as smoldering grass fires crept up the hillsides, trees exploded and burning cinders dropped from the night sky like a shower of fireflies. We discussed the local conditions, prevailing winds, the weather forecast and the night's events. Someone gave us a flashlight to take home. This was exactly the help and information I needed to safeguard my family and home, and I appreciated knowing there was a community of caring people nearby that functioned as a cohesive unit. Now, I only had Ben unaccounted for and I prayed that he had found a safe spot. In fact, I found out the next day that he and his friends were helping to put out some fires not far from us.

We stayed on the foreshore until 1 am when the wind shifted away from the west to an easterly sea breeze, and I knew then it was safe for us to go home for the night.

A knock on the door woke me up the next morning. It was my Aboriginal friend, Colleen, and her friend, Peter; the three of us were the best of mates. I invited them to the living room upstairs, while I took a short, cold shower and adjusted to the fact that without electricity there would be no coffee for my guests that morning. Without power, even the fridge would be warm, and my perishable food would soon spoil.

Colleen and Peter were already sitting out on the terrace when I joined them. There was something special about Colleen; apart from her shiny black hair and flawless brown skin she had a powerful presence, and a perfect sixth sense. I was beginning to think that every time my life was at a crossroads, an Aboriginal messenger cropped up to herald in the change.

'So what brings you here?' I asked.

'Colleen's been staying at my place during the fires', said Peter. 'She can't get home to Gosford until the Pacific Highway opens again'.

'In that case, you're both welcome to stay', I said.

Colleen smiled, settled in a chair on the patio and whiffed the air like a retriever. While I foraged in the room temperature fridge for something to eat, Peter rigged up a crackly old transistor radio in the kitchen, and we listened to the news bulletins. Lives had been lost yesterday, as Sydney battled sixty fires, extending all the way up the NSW north coast. Now most had come under control.

'Looks like the fires are just about out', I said, as I served lukewarm lemonade and tepid grapes.

'Don't be so sure', said Colleen.

'Think positive', I chirped, trying to stay optimistic.

Soon Justin appeared and joined us for breakfast. Outside, the temperature had already gone up past 35 C but it was tolerable under the shady Liquidambar tree that overhung the balcony. I looked around my tree-studded property, and realised what a mixed blessing those trees were at the time.

I was relieved when Ben phoned to tell me he was over at a friend's house, relaxing after the hectic night. His friend had an entire studio flat to himself, which was a real magnet for the peer group. He sounded excited about his fire-fighting effort the night before, and told me he'd taken some good photos of the fires.

Meanwhile, I looked forward to spending the day with my friends and was relieved the drama was over. Now life would return to normal. The only damage was to my car when the heat of the fires had bubbled the brown paint job on the aging Ford, which was ill-equipped to withstand any more punishment than it already had, over the years of non-maintenance.

Around midday I made a fruit salad from the selection in the fruit bowl. I discovered that odd things can happen in the smoke of major fires, since the bananas had gone brown and motley overnight, even though they'd been green and unripe the day before. Strangely, they'd skipped the golden-yellow stage and progressed straight into spoilage.

As we ate lunch, Colleen went quiet and listened intently. 'There's been a pressure change', she announced matter-of-factly.

'What do you mean by that?' I asked, as I ate a chunk of melon. I looked at her skeptically, glanced over to Peter, and laughed. Peter, however, didn't laugh, since he'd known Colleen for longer than I had.

'Something's coming', she said.

I looked out over the railing, and sniffed the air just as she had, but unlike Colleen, I didn't know what I was sniffing.

'I don't know what you're on about'. I told her. 'They gave us the all clear on the radio'.

Colleen couldn't be bothered arguing with me. She didn't have time for skeptics or the media, because she had absolute trust in her intuition.

'Okay', I relented. 'Tell me what you mean'.

'Look over there', Colleen said, pointing out over the balcony railing. 'What do you see?'

'I don't see anything different from five minutes ago', I said, squinting my eyes. 'Same old trees, Pittwater and Scotland Island in the distance'.

'Then, what do you hear?' Colleen persisted.

I strained to listen but heard nothing. 'Zilch' I said. 'What do you hear?'

'I hear exactly what you hear', said Colleen. 'Nothing. That's the whole point I'm trying to make'.

I was thoroughly confused for a minute, before it dawned on me that the atmosphere was eerily quiet. There were no crickets, birds or cicadas to be seen or heard. Everything with wings had disappeared, the rest had taken shelter.

'And what do you reckon about those trees over there?' asked Colleen.

After looking at them as hard as I could, I thought there was something unusual about them, but couldn't quite put my finger on it.

Colleen patiently explained it to me. 'Trees talk to each other, and the ones all around us have already closed their tops'.

Then I noticed it myself; the trees surrounding us had folded up their branches from the midsection to the crown, like the plastic Christmas trees that collapse and fit back into the box for next year.

I caught Peter listening carefully. He'd done some bush survival courses, but this was new, even to him. As I tried to come to terms with the idea of talking trees and psychic insects, my cat went into hysterics, screamed like a banshee, and let out a throaty growl before bolting downstairs and taking refuge in the bedroom. Meanwhile, Colleen was calm as a Buddhist Monk.

Just as I was getting nervous about what might happen next, the wind shifted and in blew another scorching westerly, off the hot desert centre of Australia. Not long after the furnace wind arrived, the smoldering embers from the night before erupted and set the neighbourhood alight in what seemed like spontaneous combustion. This time, trees close to us ignited with a blast that sounded like a blowtorch; while the sun disappeared behind a veil of smoke again, casting an eerie orange glow on the landscape. When a sudden gale, carrying smoke and cinder-ash, threatened to blow us off the balcony, we sprinted indoors and closed the windows to preserve whatever clean air was left in the house.

I tried comforting Justin, but I heard myself sound unconvincing, since my body was responding to the impending danger in a primal way and I was covered with goose bumps, despite the unbearable heat.

The four of us watched stunned through the window as the fire devoured trees nearby, and we listened as the young saplings exploded and the crackling flames devoured the leaves. For a few minutes, we took turns darting outside to hose down the yard, but despite wearing scarves and rags around our nose and mouth, the heat soon drove us back inside. The time to evacuate had now clearly passed; it was too risky to go anywhere, and we had to ride it out.

Just as we thought conditions could get no worse, we heard a fierce and scorching wind blow up. We heard it rattle the windows and felt its hellish heat on the inside window panes. Outside, the smoke and burning debris were obscuring visibility and building up to a gale force storm. We watched horrified as the antenna blew off my roof and disintegrated, with the aluminium sticks from the aerial flying around the backyard in a vortex, along with roofing tiles, insulation, and other people's washing, still attached to clotheslines.

We didn't know it, but we were in the middle of a fire-storm that had formed when the intensity and heat of the fire had created its own local weather system. There was nothing to do but pray for it to pass and hope for the best. My cat had known it was coming and took refuge in the safest part of the house. The insects, birds and animals had known it was coming, and had either left the area or found a refuge. Even the trees had known and folded up their branches. But we humans were usually caught unawares, and could only wet down our clothing, comfort our children and watch from the shelter of our homes. However, we were luckier than most to have Colleen, our Aboriginal Nature whisperer, who was still listening intently for signs that the rest of us had no clue about.

Even to me, there was something about this firestorm that made my hair stand on end. I was used to storms and gales out at sea and had even sailed through the heavy swell of a cyclone. But this was different and difficult to describe in words; the heat and force of this wind felt like it had blown straight out of hell.

The four of us huddled together while the fury outside continued. The temperature inside the house had climbed to 47 C (116.6 F); it was hotter still outside, and I wondered how much heat humans could withstand. Just as I toyed with the risky idea of an emergency evacuation, Colleen sensed another 'pressure change'. A few minutes later, the wind changed and the infernal force of the firestorm abated. Colleen had accurately called every shift before it happened.

A few hours later, dusk merged with the rising smoke, covering the neighbourhood in a thick, grey blanket. The night brought a total blackout again, and we lit candles inside, while outside embers glowed in the hot summer night. It was impossible to check for damage until morning.

That night in bed, I thought about what Alice had said the day before, and remembered my prayer for a change in my life. I wondered how this baptism of fire would help me and the kids live a better life. But when the drama was over and I could survey the damage to my house, I realized that all my effort to fix the outside of my life had come to nothing. It was the inside that needed fixing, and I wondered were to start.

Truth at Last

And ye shall know the truth,
And the truth shall make you free.

— St John 8:32

That February, Ben left for university, while Justin and I got on with cleaning up after the fire. The damage to my house was fixable, but the damage to my body from the poisoning was now becoming a serious problem. While I arranged for the insurance company to organize the home repairs, I was too sick to return to work. My symptoms now included sciatica, acute joint pain, as many as three excruciating migraines a week, brain fog, short-term memory loss, tingling in my hands, nausea, vomiting, dizziness, heart palpitations, anaphylaxis and increased sensitivity to an ever increasing variety of chemicals.

When I hadn't been able to work for the past two weeks and my paid sick leave was running out, I realised that without medical help for these troublesome health problems I would soon be facing financial problems. My only option was to go back to the medical centre to ask the GP for a few tests in hopes of finally finding out what was causing my illness and to have it treated. By then, I had been on a myriad of different migraine preventives and copious anti-inflammatory drugs for the joint pain. The medication did little to stop the headaches or relieve the painful joints, but I did develop a stomach ulcer from the anti-inflammatory drugs, and had to take Zantac to relieve the searing pain in my stomach.

Fortunately, on my first visit back, my GP honoured my request and ordered more tests. The x-rays revealed no abnormality, while the lumbar spine CAT scan showed two bulging lumbar discs. That explained the sciatica, but left the other symptoms still unexplained, including the tingling hands, memory loss and mind-numbing fatigue.

I returned to the medical centre to see a female GP who had recently joined the practice. By then my medical file had grown to the size of an epic novel.

'Why do I always feel exhausted?' I asked her.

She looked at me through narrowed eyes and asked, 'Are you having any trouble with your partner?'

'I don't have a partner'. I answered, before trying a different question. 'Then why are my hands tingling and why do I have problems with concentration?' I asked. 'Do you think I could be in the early stages of multiple sclerosis? What about an MRI?'

'MRIs are very expensive and there's a long waiting list', she came back with.

I'd been blocked at every turn and was now clutching at straws. 'What about alternative medicine, or are there some natural approaches that could help me?' I asked.

The doctor looked up from her notes and glared. 'Absolutely not!' she insisted. 'These things are not evidence based, and can be dangerous'.

I was alarmed at her intense response, but took her warning on board, maybe these things were unproven and I should stay away from them, I thought. By then, I knew the doctor's patience was wearing thin. Perhaps the idea that I might be a hypochondriac crossed her mind. In any case, my five minutes was well and truly over.

'Are you sure you're not having trouble with your partner?' she asked again, with her pen poised over the prescription pad. 'Perhaps you're under stress'.

When I realised that she was about to write me a prescription for antidepressants or tranquilizers, I left the surgery. The only amusing aspect to the fruitless consult was that I was barely able to do the food shopping, let alone conduct a romance, and she just didn't want to accept the fact that I had no partner. My problem was that nobody was finding out what was causing my health issues and, if they continued, I would no longer be able to work and support my family.

I returned home from the medical center feeling so frustrated that I threw out all the medications I'd been prescribed. Without the symptomatic relief these medications gave me, I ended up in bed exhausted while my joints and head ached non-stop. I barely had enough energy to get Justin off to school in the morning, or make dinner for him at night, and I rationed my energy for shopping and chores while doing only the essential things.

One day I was in bed wondering how I would be able to make the next mortgage payment, when the postman rode by on his postie bike and blew his whistle to signal that he'd delivered my mail. For the tenth day in a row, I didn't go out and get the mail from the mailbox, which by now was filled to capacity with uncollected mail. The postman must have become worried about me and placed a phone call to the local police station.

A half hour later there was a loud knock on my door. A glance through the window revealed a police cruiser parked outside my house. I wondered if Justin was in trouble, before hesitantly opening the door to a burly policeman. The officer looked me up and down before asking if I was alright. Briefly I thought about his question, which no one had ever asked me so directly before.

'No', I admitted. 'I'm not alright at all'.

In my desperate struggle to survive, it was the first time I had admitted, even to myself, that my life and health had hit rock bottom.

The police officer had only intended to check that I was alive, and once he'd established that fact, he left me to my own devices.

After closing the door, I knew it was time to make a critical decision about my life. I realised, that since no one had helped us, I had to take full responsibility for our health. And I vowed then to do whatever it took to find a way of restoring us to health and balance, and that included helping Justin achieve his full potential. This, I promised to do whatever the cost.

Once made, my decision galvanized me into action. I got dressed and drove to a second-hand bookstore at Newport, where I browsed among thousands of second hand health books, along the rows of shelves. The proprietor reminded me that there were still more books stacked at the back of the store in a small section of books that were rarely read. There, I picked up the first edition of A—Z of Chemicals in the Home, published by the Toxic and Hazardous Chemicals Committee of the Total Environment Centre in Sydney. I was glued to the spot as I read it from cover to cover, until the owner gently reminded me that it was closing time. I bought the dusty little book, along with a few other health books and hurried home to Justin who was now home from school. The book listed all the symptoms of chronic and acute chemical poisoning, and I knew instinctively that this is what had happened to us, and that this was the cause of my illness and Justin's health problems since his birth.

The very next day I phoned the Total Environment Centre, where the staff gave me the number of Dr White, a medical doctor who specialised in environmental medicine.

Dr White had a two week waiting list, but I gladly made an appointment for myself and Justin, since after eleven years of suffering, any wait would be well worthwhile if it led us to some help. The breakthrough had come just in time for Justin, who had grown into a pale and thin boy with knobby joints. The ADHD medication clearly wasn't working for him over the long term, and since taking it he'd suffered from sleeplessness, poor appetite, debilitating muscle cramps, nasty migraines, and frequent abdominal pain episodes, some requiring hospitalisations. He had also developed allergies, and his learning problems and hyperactivity had not improved.

Two weeks later, I felt slightly nervous when I walked into Dr White's clinic at Mosman, and wondered if this would be just another futile exercise as so many encounters with doctors in the past had been. When I was called in, I was pleasantly surprised to meet the quiet and thoughtful doctor, who looked to be in his late forties.

During the consultation, he not only asked dozens of probing and intelligent questions about my medical history, but he also listened carefully to my answers. It was the first time I'd had an intelligent conversation with a health professional who seemed genuinely interested in what I was saying and how I was feeling.

The doctor took a detailed medical history, and documented my symptoms in his records. He followed this up with the most complete physical examination I'd ever had from a doctor, which included palpating my abdominal organs and checking the condition of my skin, ears, throat, teeth and limbs. This approach was different from past consults when doctors zeroed in on the affected part, with pen poised over the prescription pad.

'What do you think the problem is?' I asked him, when he had finished.

'You show all the signs of chronic fatigue syndrome and multiple chemical sensitivities, for starters', he replied. 'When was the last time you felt well?'

Tears welled up when I realised that I couldn't really remember feeling well, it was so long ago. 'I'm sure it was some time before 1984', I said.

'Was that before you had first heavy exposure to pesticides?' he asked.

'Yes. Do you think that has anything to do with it?' I asked.

'Without a doubt', he said. 'But we'll do some lab tests anyway to make sure'.

Dr White ordered a battery of lab tests, including standard haematology and biochemistry and additional blood tests to determine the presence of organo-phosphate and organochlorine pesticides. Then a sample of my hair was sent to a lab, in the USA, to determine the presence of heavy metals and the levels of essential minerals in my body. He ordered another test to determine what degree of sensitivity I had to the toxic substances I'd been heavily exposed to. It would be a month before the results were available.

The sudden relief I felt at having this eleven year nightmare validated by an intelligent, caring human being was something beyond mere relief. I felt liberated knowing that I wasn't lazy or crazy; that Justin and I got sick after we were chemically poisoned.

'Why haven't we been able to get any help over the past eleven years?' I asked.

'You fell through the cracks', said Dr White.

'Why is that?' I asked.

'It happens all the time to people with environmental illnesses, said the doctor. 'Medical students aren't taught about the damaging effects of chemical exposure because the drug and chemical companies have a huge influence on medical education...you figure it out'.

I was shocked to hear it, but after what we'd been through, it wasn't hard to figure out, or believe. Now I knew why no doctor, no hospital and no paediatrician could help Justin and me, from the very first time we were poisoned. They weren't educated about chemical exposure because the industry wanted to protect their products. That realisation would make me angry each time I thought about how long my son had suffered unnecessarily.

Dr White went on to explain that the sudden onset of anaphylaxis after the heavy chemical exposure was because my immune system had been damaged by the toxic chemicals. This damage also caused me to develop hypersensitivity to a variety of different chemicals over time and react to them, even at normally tolerated doses. He explained that a typical feature of chemical sensitization was its spreading effect, which caused sensitivity reactions, not only to the triggering chemical, but also to many other unrelated chemicals. That accounted for my severe allergies to synthetic food additives and preservatives, my intolerance to alcohol, and also to my increasing sensitivity to plastics, cleaning agents, paints, solvents, fragrances, and chemicals in personal care products. He also explained that my heart rhythm irregularities were a common reaction to these chemicals.

While we waited for the test results to come back, Dr White said he would start me on nutritional supplements to improve my liver function, raise my nutritional status, and strengthen my immune system. As my liver function improved, he said, some form of detoxification could begin. At my next appointment, he suggested I bring Justin in for his first consultation.

When I walked out of the doctor's office, I knew that a corner had been turned in my life. The pieces of the puzzle were rapidly falling into place and the changes I had hoped for seemed just around the corner. I was so relieved, that I cried all the way home.

As soon as I walked through the door, I mixed the nutrient powder the doctor had given me in the blender, with a ripe banana and soy milk, to make a smoothie. After downing it, I glanced at the open cupboard under the sink where I had dozens of chemical products stored, and remembered what the doctor had said about limiting unnecessary chemical exposures. I discovered flea powders and flea rinses, cat collars, fly-sprays, surface sprays, oven cleaners, carpet deodorisers and harsh detergents.

From there I went to the bathroom cupboard where I found dozens more chemical products such as: personal insect repellent, head-lice treatments, air fresheners, perfumes and chemically laden personal care products. Even the toothpaste had a health warning on the label. Meanwhile, in my laundry and garage I discovered highly fragranced detergents, pesticides, herbicides, solvents, paints, weed wands, aphid powder, and surface sprays for killing crawling insects.

When I discovered, with the help of the A-Z of Chemicals in the Home, that many of these products contained a variety of toxic chemicals, including the same organophosphate chemicals that had poisoned us in the past, I went outside to get my empty rubbish bin.

For the next hour I filled the bin with the chemical products and left only mild detergents and the vinegar and bicarb combination that my mother used to recommend for cleaning (I didn't know yet that there were non-toxic products available). By the time Justin came home from school that day, I had a full bin and the satisfaction of knowing that my home was no longer a source of chemical pollution.

The next day I decided to explore some natural therapies, against the GP's advice, and made an appointment with a massage therapist who gave me a relaxing foot treatment (reflexology). That night, I had my best night's sleep in years.

After I'd cleared my house of harmful chemical products and after the relaxation treatment, I realized that I hadn't had a migraine in three days, and was able to return to work the next day, just in time to avert serious financial problems after my paid sick leave had run out.

The patient on my next work shift was deeply unconscious after sustaining a severe brain injury in a car accident. In fact, he was in such a deep coma that he could not regulate his own breathing and had to be on a ventilator for life support. Beeping on a shelf above his bed was a monitor with digital readouts of his vital signs, including his blood pressure and the central venous pressure inside his body. My job was to regulate several drug infusions that would keep his vital signs within normal limits, but due to his brain injury he was constantly hypertensive, and I had to up-regulate one of the infusions to keep his blood pressure within safe limits. Despite the higher drug dose, however, he remained hypertensive. After a few hours of this, I decided to administer pressure area care and to shift his position. Thinking that it would make no real difference, since the man was deeply unconscious, I also gave him the same type of foot massage that had done me so much good the previous week.

Shortly after finishing, I glanced at the monitor and realised that the unconscious man's blood pressure readings had come down into the normal parameters, and were staying there. I was astonished to realise that this simple type of natural therapy was a safe and powerful healing tool – and there was no question that it could have been a placebo effect on an unconscious patient

That afternoon, on my way home, I understood for the first time, what Colleen and the Indigenous culture had been trying to teach me: there was a life force in the body and a consciousness (in all living things) that operated outside the normal waking state. The Chinese also knew of this life force and called it chi. They used it in acupuncture, which stimulated the energy points in the body, to heal and balance body energy.

The feet had many such energy points. I realized that Western medicine had a lot to learn, and thought that I should begin exploring the healing possibilities outside the rigid conventional medical system. I was also determined to find the science behind natural medicine.

A few weeks later I went back to Dr White, this time with Justin. My test results showed that repeated chemical exposures had depleted my body of the essential minerals, vitamins and amino acids that were required to maintain all body functions, but especially the Cytochrome P450 pathways in the liver that detoxified environmental chemicals. These pathways had already been damaged by the initial poisoning, and their breakdown led to a buildup of the chemicals and heavy metals in my body, at levels that were making me sick. Leaving my condition untreated would mean I was at risk of cancer, degenerative diseases and early death.

Dr White ran similar tests on Justin, and when they came back we were in for another surprise. His nutritional levels had also been depleted by the chemical load, and apart from the toxic residues he was also carrying, the test showed he had extremely elevated lead levels in his body — seven times the amount known to cause behavioural and learning difficulties.

It was not surprising that he couldn't concentrate, and amazing that he had managed school, at all, up to now. Apart from lead, he showed elevated levels of aluminium, cadmium, mercury, and a small amount of uranium. Justin's liver function had also been damaged by the heavy exposure to pesticides, and his impaired detoxification pathways caused an accumulation of chemical residues and heavy metals to build up to toxic levels. This, in turn, affected his brain function.

Dr White suggested nutritional supplements and a course of intra venous (IV) chelation to remove the lead from Justin's tissues, and to follow it up with powerful antioxidants, IV vitamin C and glutathione to repair the chemical damage.

These treatments meant that I would have to take Justin to the clinic each week and stay with him for 4 hours while he received the IV. The ongoing treatments meant a significant expense, for which I would need to find the funds; but they were the first real options we had ever been offered to treat Justin's condition. At the time, I was already stretched beyond my financial limits with the mortgage, the loan and the cost of childcare and, just as I was trying to figure out how I was going to afford Justin's treatments, Dr White hit me with another curveball.

He told me that if I wanted a chance to recover from the organ damaging effects of this serious poisoning I would need to take a sabbatical from work, move to an environmentally clean area, and wholeheartedly undertake treatment. Even then, there were no guarantees the treatment would work. In fact persons with such advanced environmental illness, he said, very rarely recovered. All this came as a shock; firstly that I had a largely incurable condition, and secondly, I couldn't imagine quitting my job and without guarantees, it seemed like he was asking me to jump off a cliff and hope for the best. In any case,

Dr White had given me the information, now I had to evaluate it and decide what to do.

In the car on the way home, Justin asked me when he could start the treatments. I wasn't expecting this from him, since he'd resisted taking his ADHD medication and wasn't crazy about seeing the paediatrician who had prescribed it. But his innocent willingness to get well reminded me of the promise I'd made not so long ago, that I would find a way to help him, no matter what the cost.

The next day I phoned Dr White's clinic and scheduled Justin's fist IV treatment to start the process of clearing the lead from his system. While waiting for it, I started Justin on the doctor's supplements and introduced organic food into our diet to further reduce exposures to chemical residue.

As to my own treatment; I was still undecided about making the changes the doctor had suggested, and before taking on such huge life changing decisions I wanted to see how Justin would go with his first chelation treatment.

Two weeks later, Justin was sitting in Dr White's intra venous clinic room with ten other patients who were having various other types of IV's. Inserting the cannula had been traumatic for him, but once the IV started, he settled down and chatted happily with the other patients in the room.

Justin spent the next week without headaches and I didn't hear from Justin's school once. He came home from school one day with pink cheeks instead of his usual pale face.

'Hey Mum', he said, 'I think I'm feeling better'.

That was just the sign I needed. Now that Justin was making progress, I could risk selling my house and relocating to an environmentally clean place where we could focus on recovering from the chemical injury. The sale would also relieve me of debt and allow me to take the sabbatical from work, and we would use the proceeds to finance the medical treatments. Hopefully, after a while, I would be well enough to return to work.

I looked at my boy, who still stood there grinning at me. 'I think you're beginning to get some colour in your face' I observed.

Justin nodded. 'By the way, Mum', he added. 'Could I have a dog?'

I hesitated, 'We'll see', I said, not thinking it was a good idea, just before we made a major move.

'But I would love a dog and he would love me', came the reply.

On my next day off work, a friend came over to visit. Lois was the mother of Justin's friend, one of the boys he'd met at soccer.

'Come with me to the animal shelter, and help me pick out a dog', Lois said over a coffee.

I looked at her apprehensively. 'I don't think I should go anywhere near an animal shelter', I answered. 'You know what I'm like'.

'Don't be silly', she said. 'Just come along and help me look'.

The next day we set out on a long trip across town to Yagoona, where the RSPCA had an animal shelter. The place was scrupulously clean, the staff was fantastic, but the sight of so many homeless and abandoned animals stirred both of us up into an emotional state.

While Lois looked at a female German Shepherd, I wanted to avoid any temptation to get an animal, by walking away. But in a nearby enclosure, an Australian Cattle Dog caught my attention. She was a sturdy, medium size dog with a white and red ticked coat, and a tan patch over one eye.

The breed is the world's foremost cattle herder and was bred in the Australian Colonial era for the dangerous job of mustering and moving large herds of cattle in the toughest environment of the Australian outback.

On our Mudgee property I had seen these courageous dogs muster up thundering herds of cattle, while the men waited for the dogs to do the perilous work, before droving the herd on horseback to their destination. I have seen cattle dogs disappear into a galloping herd of cattle, and emerge out the other side, to dust themselves off. They will take on the most bad tempered herd bulls by latching on to their heels with their strong jaws, until the 2500 lb (1134 kg) animal surrenders and heads in the direction the dog wants him to go.

I knew these dogs needed a strong pack leader, but once they respect their human, they are loyal to the end, and will lay down their life for their owner without hesitation. Australian bush stories abound of cattle dogs that have waited, in vain, by their owners' graves.

Once I had seen this intelligent and loyal breed in action on the land, my heart was forever theirs. Now a cattle dog was sitting on her haunches and waiting to greet me as I walked by, and I could not stop myself from looking at her. The card on the wire mesh fence told me her name was 'Emma'.

'Hello Emma', I said, half expecting her to look away, but she looked straight up at me as if waiting to see what directions I would give her next. When she maintained eye contact with her intelligent brown eyes, I bent down to get a better look. She was a young adult, in need of a good feed, a job and a boy to love her.

'I suppose you want to go home with me?' I said to her.

She cocked her head to the side before lying down in front of me, as if to say, 'I'm at your service. What took you so long?'

It wasn't long before Lois took home Sheba, a German shepherd, and I took home Emma the Heeler.

When Justin came home from school that day, it was instant love between boy and dog. Emma sniffed around the house for a while, and after a big feed she settled at Justin's feet as though she'd always been with us. Emma sensed that her job was to be a companion for Justin, and that was exactly what I had wanted her to do. She knew when Justin was due home from school and waited at the front door for him every day at a quarter past three. The two then played happily all afternoon, and they were the best therapy for each other.

Now that we had a dog, it was time to put the house up for sale, quit my job and find a sanctuary.

Time to Heal

The wound is the place where the Light enters you.

— Rumi

I found the ideal property for us in the high country, near Canberra. The five acre spread came with a four bedroom homestead. There was plenty of room for Justin and me, and an extra bedroom reserved just for Ben when he came home during university holiday breaks. From our living room window we enjoyed looking at the snow-topped Brindabella Ranges.

The cold, dry climate made the warm fireplace the social centre of our home, where we ate meals made from our home-grown vegetables, and enjoyed the friends and neighbours that dropped by.

It wasn't long before Ben arrived home on his holidays and loved spending time with us in our quiet sanctuary, over Easter and Christmas. He was doing well at university and planning to study in China, for a year, to hone his linguistic skills in Mandarin.

Emma was in her element when both boys were home, and had fun playing ball games with them. Her job on the property was to keep it clear of rabbits in winter and free of snakes in summer, and I always felt safe when she was around.

Our acreage gave us everything we needed, including firewood, which Justin helped me to split and stack, ready to use in the fireplace.

To help us recover from the toxic overload of the past years, we maintained a clean and green environment to lower our daily chemical load. This we did by gathering rainwater into a large tank and using an EnviroCycle septic system for wastewater, a system that worked only without harsh detergents.

The bicarb and vinegar I had been using to clean the house with was working well, but I missed my makeup and personal care products, until someone put me onto a line of non-toxic personal care and household products that were not damaging to our health. Once again, I could occasionally wear makeup, wash my hair with non toxic shampoos, and my household and washing were clean and white, and non-polluting to the environment. I was beginning to think my grandmother had been right all those years ago, when she warned about going against Nature. Now that we were more aligned with Nature, our lifestyle was more balanced, our lives were happier and our health was starting to improve.

From the unpredictable grind of my shift work, we now had a regular routine. Each morning I drove Justin to the main road to catch the school bus that took him to a local Canberra school. Each afternoon I picked him up with Emma, who went wild when she saw him step off the bus. And every evening we enjoyed our home and sanctuary.

Once a month, Justin had to have a few days off school when we returned to Sydney for another IV treatment at Dr White's clinic.

Often, the doctor administered an IV to me as well, consisting of either vitamin C or glutathione, the natural antioxidant that was most effective in protecting humans from the effects of chemical exposure, as well as repairing tissue damage from chemical toxicity (Smith 1996).

During the two days that followed the IV treatments I felt full of energy - the way I'd felt before I was poisoned - but it would take many more treatments before I would experience a more lasting improvement.

Meanwhile, the doctor ordered some further tests for food sensitivities, and after discovering that both Justin and I were highly (immunoglobulin G - IgG) reactive to wheat gluten and cow's milk dairy products, I learnt how to cook wheat and dairy free cuisine, using delicious substitutes for those foods.

With these dietary changes, and nutritional supplements, Justin developed a healthy glow and looked more robust. On his next appointment, Dr White also discovered that Justin had grown several centimeters and gained enough weight to put him into the 70 percentile on the development chart, since stopping the ADHD medication.

I too was beginning to look healthier; my focus and concentration were better, my hair was thicker with less grey and my nails were stronger. I hadn't had an asthma attack in six months. With these early signs of improvement, I realised that if my condition had been left untreated, I would not have been able to continue working, and was on track to suffering major illness, disability or death.

Help came just in time for both of us, and with better health, our lives started changing in other areas too. Friends gradually came back into our lives and Bill started phoning to chat with me about Justin's progress. We reconnected as parents to the boys, and also as friends. He phoned me regularly each week for the remainder of his life, and we became better friends than we had ever been spouses, and shared a lifelong rapport and mutual respect for each other. He re-bonded with the boys too, as far as distance allowed. Justin rode the Greyhound bus to Sydney one weekend each month to see Bill, while Ben kept in touch with us by phone. He had now been approved for the educational exchange program, and was on his way to China where he would live and study for the next year.

Before leaving for Asia, Ben spent some time at home where, together, we made a plan to travel to the US to reunite with my family. Finally, after many years, the boys could spend time with their grandparents, aunt and cousins. This long overdue homecoming and reunion was possible only because I had funds from the sale of the house. These proceeds also made it possible for me to help Justin in a way that I'd never been able to do before. After Justin had suffered unnecessarily for so many years, I was glad we had found Dr White, but I also wanted a specialist's diagnosis from the top medical specialist in the world, and that meant traveling to Los Angeles. From there, Justin and I would fly east to visit my family, and Ben would join us there later.

The 16 hour flight to Los Angeles was exhausting enough, but after clearing customs, Justin and I lugged our suitcases outside the terminal building, only to hail a cab, whose driver would give us the ride of our lives. The young guy talked as fast as he drove, and seemed intoxicated on some kind of stimulant while speeding through the LA traffic like a Kamikaze. Justin and I were both rigid with terror in the back seat and relieved when the driver screeched to a halt in front of our hotel.

After sleeping off some of the jetlag, we set out for our appointment with the medical specialist the next morning. Dr Heuser was a tall, bear-like man who put us at ease immediately. Apart from his warmth and sincerity, he was also a highly qualified MD, a medical specialist, a PhD researcher and author of several studies on chemical injury in the world's medical literature. After welcoming us, he took a thorough history and ordered a battery of blood tests, respiratory function test and high tech brain imaging scans.

After telling Dr Heuser about the wired taxi driver, he suggested we use his clinic's driver who was designated to take overseas visitors to where they needed to go for the medical tests. The next morning Bob, the driver, picked us up from the hotel and insisted we accompany him to Denny's first. There Bob had us in stitches with his stories, while he devoured a cholesterol laden breakfast of eggs, bacon and hash brown fried potatoes. Only then, was Bob ready to deliver us to UCLA Medical Center, at Irvine, for a special brain (SPECT) scan.

After a grueling couple of days of tests, Justin and I sat in Dr Heuser's office for the verdict. My blood tests showed several abnormal autoimmune reactions, similar to lupus, and my SPECT scan had shown abnormalities in brain blood perfusion, a typical legacy of pesticide poisoning. Dr Heuser confirmed that Justin had sustained a brain injury from his prenatal exposure to organo-phosphate pesticides. He confirmed that this chemical injury had caused learning difficulties, problems with hyperactivity and impulse control, as well as the physical health problems that he had been suffering from, including abnormal liver function.

Now, after twelve years, we could finally put a name and a cause to the nightmare that had ruined our health, destroyed the quality of life for all of us - including Ben - and shattered our family. My intuition had been right all along, we had been poisoned, and suffered from its devastating effects, while people judged us and doctors offered no help. Even the paediatrician's diagnosis of ADHD was only a symptom of Justin's brain injury, and not a diagnosis in itself.

We could finally move ahead and I would continue to do whatever necessary to help Justin, and all of us, to recover as fully as possible. Now, I also wanted my own life back, and I was determined to gather my family together again after the many years of loss and isolation this poisoning had caused us.

When I asked why my son was made to suffer needlessly for 12 years of his life, Dr Heuser echoed Dr White's answer when he said: 'the industry has trouble accepting that their products make a great number of people sick'.

I wondered how many other children were sick and suffering because of unsafe chemical products, and because doctors get their education from an industry that protects its products. Asking that question gave birth to the activist inside me. I had never been an advocate for anything in my life before, but now I was determined to do something about this issue when I got back home.

After finishing our stop in LA, Justin and I flew east for the family reunion, where he met his extended American family for the first time. Ben joined us later from China and had fun reconnecting with his cousins, who were also in their twenties. For the first time in many years my parents had their entire family around them, and we experienced being in the family fold. It was truly a homecoming that gave us back our sense of family unity.

The excitement and jetlag were enough to crash Justin and me for several weeks after arriving home from the US. It made me realize that we still had a long way to go before we could resume a normal life. Just as I wondered what to do about it, the woman who managed at the local post office told me about some supplements that had helped her. I laughed and told her that I had a cupboard full of supplement pills, powders and liquids, recommended by my integrative medicine doctor, and there wasn't a supplement I hadn't taken. She probably thought that I was a tough nut to crack, and when I next went into the post office, she invited me to a meeting where, she said, I would be invited to share my healing journey. Since I welcomed the chance to talk about chemical injury, I accepted.

The meeting was in an old pub at Yass, and the smoky room irritated my eyes, but I stayed anyway to listen to the speakers. A woman about my age got up to share the story of her battle with Lupus. When she said she was a nurse, I paid special attention to what she had to say. Like me, she also had dozens of supplements on her window sill, and had made no real progress until she listened to a tape on nutrition and realized that she needed all of the 90 essential nutrients on a daily basis. This included 14 vitamins, 20 amino acids, 3 essential fatty acids, antioxidants and minerals. Instead, she was taking single vitamins and minerals in a dozen separate bottles and missing out on the remaining essential nutrients the body needs to function and heal from illness. As soon as she took in all 90 essential nutrients in nutrient dense supplements, she said, her health began to improve.

After the talk, I couldn't get my hands on those supplements quickly enough. And when I did, I took only a small amount, since I was hyper-sensitive and had to introduce new things carefully, but the difference was immediately apparent.

A few days later, I went to my local shopping centre for a few groceries. I usually struggled up the stairs to the supermarket and, at the top of the flight, I would normally have to stop and catch my breath, or even sit down on the floor until I could continue further. This day, I climbed the stairs and kept going. I was at the other end of the first floor, before realizing that I had not stopped, my limbs did not feel leaden, and I was not out of breath.

For the first time, my body was starting to effectively repair the damage with all 90 essential nutrients, instead of taking a vitamin, a mineral or an antioxidant in separate doses, as I had been doing. I decided to gradually increase my dose of 90 essentials to the allowable maximum on the label and keep this up on a daily basis. Over time, I coined a new name for this process: in intensive care we used the term 'loading' which referred to starting and maintaining a dose of medication that would result in a noticeable and beneficial effect. I now called taking all 90 essential nutrients on a daily basis: 'nutritional loading'.

After I had looked into the scientific research behind nutritional medicine, I learnt that by replacing all the life giving nutrients that were depleted through chronic illness, the body could heal itself of many different conditions, since all human metabolic processes relied on essential nutrients to function. I hadn't realised, until then, that the liver also needs all the essential nutrients daily to detoxify and clear environmental chemicals and medications from the body. This essential nutrition applies to everyone living in the chemical laden environment of the modern era, and not just those who have suffered acute chemical exposures.

By continuing this nutritional loading regimen on a daily basis, I was providing the essential nutritional building blocks for my body to repair itself at a cellular level, and when I started Justin on this regimen, his recovery also accelerated. Six months later, for the first time in his life, Justin brought home B's and an occasional A on his school report card. From then on, he started talking about his future, and it was the first time in his life that he dared look forward to accomplishing a goal.

Time to Stand Up

Be the change you want to see in the world

- *Mahatma Gandhi*

Now that two doctors, including a top international medical specialist, had confirmed that Justin and I had suffered the devastating health effects of chemical poisoning, I decided it was time to let the authorities know that a chemical product that the Australian regulator had allowed on the market, was capable of destroying the lives of the consumers who buy it and use it as directed.

In a letter to the regulator, I outlined our experience of being poisoned by the sheep-dip, Justin's brain injury and the terrible years of suffering it had caused him, the family's confusion and breakup, the crushing medical expenses and the years that Ben had been deprived of a mother's full attention while I was struggling with my own ill health and trying to help his sick brother. I didn't want this to happen to any other family and thought it was my duty to write to the relevant authority so steps would be taken to ensure this didn't happen again.

I waited several months for a reply, and when it came, I had to sit down on a chair to finish reading it. The bureaucrat, who authored the letter wrote it in cold blood, as he systematically whitewashed each of my points. In a final blow, he dismissed the (now established) fact that this chemical product had caused serious illness, as being: 'the result of a strong interaction with personal belief factors'. When he alleged that I had imagined it all, I started shaking with outrage. I had been confronted, up close and personal, with the unpalatable proof on paper that the regulator was protecting the interests of the chemical companies and didn't care about what had happened to my son or about the community's health and welfare.

That day, my world view changed permanently when I was forced to accept that we, the consumers, could not trust our own government to protect us from harmful chemical products, and from now on we would need to take personal responsibility and choose safe products from reputable manufactures. When I wrote a letter back to ask, among other things, what kind of testing the regulator conducts before approving chemical products for release on the market, I got no reply.

Instead, I was chosen as one of the participants in the community consultation process that occurred before a chemical company could license a product with the regulator. This meant that the regulator would send out data to me and others about the chemical product awaiting a license, and hopefully authorities would take the community's feedback into account before issuing the chemical company a license to release its product on the market. This way, I thought I could help to safeguard the community, if the chemical seemed harmful.

My first consultation came when a folder arrived from the regulator containing laboratory data from animal toxicity tests conducted on a pesticide that was awaiting registration in Australia.

That night I carefully read the study, which had been designed to find out the pesticide's health effects on rabbits. After pregnant females were given small doses of the pesticide, some gave birth to stillborn kittens, while other kittens were sick, abnormal or died shortly after birth. I showed this study to a scientist friend of mine, who agreed that the pesticide had caused some seriously adverse health effects on the animals, and would probably have the same adverse effects on humans if it were licensed and used in consumer products. Together, we wrote a thorough report reiterating the toxic effects on the experimental animals and strongly advised against the product's licensing, based on its reproductive toxicity.

Six weeks later I received a form letter thanking me for participating in the community consultation and advising me that the product had been licensed. I knew then that the entire community consultation process was a sham and the regulator would rubber stamp all chemical and pharmaceutical products for the big pharmaceutical and chemical companies, without any independent testing or regard for human health risks. It was shocking to discover, all over again, that our children's health was no consideration when it came to profits.

By now, it was becoming obvious that we were not the only ones who had been poisoned by toxic chemicals in consumer products. There were millions of others. One website tracks the statistics: www.healthline.com

Chemical poisoning is a major public health concern. Approximately 95% of all accidental or intentional poisonings are due to chemicals. Nearly 90% of these cases occur at home. The smallest children, infants and toddlers, are at the highest risk for accidental (acute) poisoning. In 2000, poison control centers received well over a million calls about poison exposures to children younger than age 6. Chronic exposure is chemical poisoning that occurs slowly and insidiously over a prolonged period of time. Many chronic, degenerative diseases have been linked to environmental pollution or poisoning. The list may include cancer, memory loss, allergies, multiple chemical sensitivity, chronic fatigue syndrome, infertility in adults, learning and behavioral disorders, developmental abnormalities, and birth defects in children. [5]

Around this time, I wondered how many other unregulated hazards were on the market that we were not told about. In 1996, the biotech industry was in the process of applying, to the regulator, for the first licenses to introduce and grow genetically engineered crops in Australia. By now, I had the internet and

---

[5] http://www.healthline.com/galecontent/chemical-poisoning#description

used it as a research tool to find data on the human health effects of GMOs (genetically modified organisms). I soon discovered the expert work of Dr Árpád Pusztai, and learnt of his experiments showing that rats had suffered shocking intestinal and immune system damage after being fed GMOs. For a while, I was in e-mail contact with Dr Pusztai, who was persecuted for his discovery, and fired from his position at Rowett Research Institute in Aberdeen, Scotland. Like me, he also discovered, first hand, the power of corporate interests over the interests of the community, and was quoted as saying: 'I find it's very unfair to use our fellow citizens as guinea pigs'.

After Dr Pusztai's findings, every subsequent study on animals fed GMOs found that they were damaged, rendered infertile, or they died in significant numbers - and that all animals, if given a choice, would not even eat genetically modified food.

Throughout 1996 I was frustrated by the twin issues of chemical toxicity and GMOs, as I struggled to find a way to safeguard my community from the adversity that had happened to us. At that juncture, I decided to write a book and tell our story, in hopes it would warn people of the risks of unsafe chemical (and food) products, and encourage others to take the responsibility for their health and home choices into their own hands. I figured that if I could do it, everyone could address their health problems, clean up their homes and change their lives, by making informed choices.

Over the next eight months I wrote the early version of this book and submitted a copy of the manuscript to a Melbourne publisher and a large US publisher. I knew that writers' first books were rarely published and held out little hope, but within a few weeks the Melbourne publisher offered me a publishing contract, and after asking a lawyer friend look at it, I signed it.

The very next day I got a phone call from the US: it was the acquisitions editor. The publisher that made Neale Donald Walsch a household name with the publication of his book, *Conversations With God*, now wanted to do the same for me.

For a moment, I wanted to savour the dream and imagine what it would have been like to have my message internationally known, and to experience financial security, before admitting to the acquisition editor that I had already signed a contract with a Melbourne publisher.

Christmas and New Year came and went while the Melbourne publisher prepared the book for publication, and in early 1997, my book was released, and the frantic promotional tour was on.

With a publicist, I flew from city to city and was interviewed on radio and TV; from Kerri-Anne Kennerley's Midday Show, to channel 7's 11 AM Show, to Jon Faine at ABC radio in Melbourne, and every major media gig in between. The publicity must have reached widely, since many people I hadn't seen for years phoned me to tell me they'd seen me on TV. The book went on to become a bestseller and was credited with changing consumer choices to non-toxic personal care and household products.

When the tour was over, I thought I had a chance of doing some good with the book in the corridors of power. I assumed that if only the politicians were informed about the damage that chemicals were doing to people, they would tighten the legislation and make the country safer, at least for children.

In the days when citizens still had access to their senators and MP's inside Parliament House, I took a stack of books with me and delivered them to representatives from both sides of politics, including the Greens. I thought the newly elected Greens Senator, Bob Brown, would be interested in the harmful effects of chemicals on humans, since he was a medical doctor, but unfortunately his focus was solely on tree conservation. Likewise, none of the other politicians in power were remotely interested in doing anything about the issue.

I did, however, start getting invitations to be a key note speaker in hundreds of seminars over the next few years, and it was during a flight home from one such function that I bumped into the former Attorney General who was the current Deputy Leader of the Opposition. Since we were both waiting for our luggage to appear on the carousel, I had an opportunity to speak with him for 10 minutes, and to give him a book. When we parted he made a joke about being 'sensitive to the issue' and that repartee proved the end of his interest. After retiring from politics, he joined a Brussels based organization founded by a past president of the World Bank.

Just when I asked myself why no politician would want to stop people being chemically poisoned, I was reminded of Gandhi's quote:

If we could change ourselves, the tendencies in the world would also change. As a man changes his own nature, so does the attitude of the world change towards him. We need not wait to see what others do'.

This philosophy proved to be the right one for me, as I discovered how my unique skills and abilities could lead to a new direction in my life's work. The nutritional loading regimen continued to bring spectacular improvements to Justin's health and my to own. It restored my focus and concentration, allowing me to start writing on health topics, often breaking news stories before the major media, which led to a career in freelance journalism.

By 1999, my energy levels were sufficiently restored for me to establish the first environmental health consultancy in Australia, where I showed people how to convert their homes to healthy homes and helped them to improve their own health in the bargain.

My gradual recovery also allowed me to return to university to study for a Bachelor of Health Science Degree in Complementary Medicine. This science based degree revealed the extensive scientific evidence base behind complementary medicine, contrary to what my GP had said to me a few years earlier.

Three years later, I co-founded, with a medical partner, the first integrative medicine clinic in my community, where we offered thousands of patients the best of conventional medicine and the best of complementary medicine, tailored to their individual needs.

In 2003, a tough challenge confronted me when a family came to our clinic with their 11 year old daughter, Sarah, who had been falsely diagnosed as being pregnant, when she really had a rare and deadly tumour. Due to the extreme circumstances the family found itself in, and the mistreatment that Sarah was receiving, I felt I had no other option than to be a whistle-blower. The full story is told in my book, *Sarah's Last Wish.* I had never been a whistle blower before, and wasn't fully prepared for the huge personal and professional price it would cost me. It took several years to recover from the fallout from this action, but despite some grueling years with heartbreaking losses, I have never regretted this decision. In fact, these years have brought their own gifts: both my books achieved the best sellers list and I found the time to pursue a master's degree and a PhD.

Though our early journey was full of obstacles and perplexity, it also gave us knowledge, experience and resilience. Ultimately, overcoming the challenges united us as a family.

Justin continued his treatment, and was gradually able to make up all the gaps in knowledge that his injury had robbed him of. Over several years, he continued to doggedly pursue further education and additional qualifications until he was able start his chosen vocation. He now enjoys a career as an English teacher and is married to a lovely young woman who is studying to be a nurse. His compassion for animals has led him to rescue several strays, which now enjoy a place in his home and heart.

Ben has grown up to be a principled and gifted man in his 40's. After earning two university degrees, he works as a legal professional with a large Sydney firm. He married a beautiful and talented woman, who works as a cancer researcher and lecturer. They have a delightful and healthy daughter.

The passing years have brought me to a place of deep gratitude to the Creator, who has granted all my wishes, saved our lives, softened our hearts, and guided my life's work.

I am reminded each day that it was only through our love for each other, and a lot of determination, prayer, and nutritional medicine that we were able to live the best lives we could live.

Time to Stand Up

About the Author

Eve Hillary BHSc. MA, ND is a health practitioner in complementary medicine, a health journalist, and best selling author.

She co-founded the first integrative medical clinic in her community, offering patients the best of conventional medicine and complementary medicine. Her current practice is in nutritional medicine and sports nutrition. Her clients include elite ultramarathon runners and other endurance athletes.

Eve has authored *Sarah's Last Wish*, an unforgettable story of a courageous teen-aged girl who stood up for her right to dignity and more medical choices during her terminal illness. Its message is creating an awakening to better health care.

Her latest book, *Beyond the Toxic Harvest*, traces the healing journey of her chemically injured family. Its powerful message is creating awareness about environmental health, safe consumer choices, and how to achieve wellness and maintain a healthy balance in a toxic world.

As a health journalist, Eve writes and speaks on past, current and future trends in healthcare.

Contact the author: sarahslastwish@gmail.com

Ojibway Prayer

To Walk in the Sacred Way
Grandfather,
Look at our brokenness.
We know that in all Creation
Only the human family
Has strayed from the sacred way.
We know that we are the ones
Who are divided,
And we are the ones
Who must come back together
To walk in the Sacred Way.
Grandfather,
Sacred One,
Teach us love, compassion, honour
That we may heal the earth
And heal each other.

References

Abel-Rahman A, Abou-Donia S, El-Masry E, Shetty A, Abou-Donia M (2004) Stress and combined exposure to low doses of pyridostigmine bromide, DEET, and permethrin produce neurochemical and neuropathological alteration in cerebral cortex, hippocampus, and cerebellum. Journal of Toxicology and Environmental Health Part A 67: 163-92.

Adgate JL, Barr DB, Clayton CA, Eberly LE, Freeman NC, Lioy PJ, Needham LL, Pellizzari ED, Quackenboss JJ, Roy A, et al. Measurement of children's exposure to pesticides: analysis of urinary metabolite levels in a probability-based sample. Environ Health Perspect. 2001 Jun;109(6):583–590

Albright JF and Goldstein RA (1992) Is there evidence of an immunologic basis for MCS? Toxicol Ind Health 8: 215-219

Aldridge JE, Seidler FJ, Slotkin TA (2004) Developmental exposure to chlorpyrifos elicits sex-selective alterations of serotonergic synaptic function in adulthood: Critical periods and regional selectivity for effects on the serotonin transporter, receptor subtypes, and cell signaling. Environ Health Perspect 112:148–155.

Altenkirch H (2000) Multiple chemical sensitivity (MCS) - differential diagnosis in clinical neurotoxicology: a German perspective. Neurotoxicology 21: 589-97

American College of Occupational and Environmental Medicine (1999) Position statement. Multiple chemical sensitivities: idiopathic environmental intolerance. J Occup. Environ. Med. 41: 940-942

American Lung Association, Environmental Protection Agency, Consumer Product Safety Commission and American Medical Association (2009) Indoor Air Pollution: An Introduction for Health Professionals.

American Medical Association Council on Scientific Affairs (1992) Clinical ecology. JAMA 268: 3465-3467 - 82 - A Review of Multiple Chemical Sensitivity

Anderson RC and Anderson JH (1999) Sensory irritation and multiple chemical sensitivity. Toxicol Ind Health 15: 339-345

Andersson MJE, Andersson L, Bende M, Millqvist E and Nordin S (2009) The Idiopathic Environmental Intolerance Symptom Inventory: Development, evaluation and application. J. Occup. Environ. Med. 51: 838-847

Antelman SM (1994) Time dependent sensitisation in animals: a possible model of multiple Chemical Sensitivity in Humans. Toxicol Ind Health 10: 335-341

Arnetz BB (1999) Model development and research vision for the future of multiple chemical sensitivity. Scand J Work Environ Health 25: 569-573

Ashford NA (1999) Low-level chemical sensitivity: implications for research and social policy. Toxicol. Ind. Health 15: 421-427

Ashford NA and Miller CS (1991) Chemical Exposures: Low Levels and High Stakes. New York: Van Nostrand Reinhold.

Ashford NA and Miller CS (1998) Chemical Exposures: Low Levels and High Stakes (2nd edition). New York: Van Nostrand Reinhold.

Baines CJ, McKeown-Eyssen GE, Riley N, Cole DEC, Marshall L and Jazmaji V (2004) Case-control study of multiple chemical sensitivity, comparing haematology, biochemistry, vitamins and serum volatile organic compound measures. Occupational Med. 54: 408-418

Baines CJ, McKeown-Eyssen GE, Riley N, Marshall L and Jazmaji V (2007) University of Toronto case-control study of multiple chemical sensitivity-3: intra-erythrocytic mineral levels. Occupational Med. 57: 137-140

Bartha L, Baumzweiger W, Buscher DS, Callender T, Dahl KA, et al. (1999) Multiple chemical sensitivity: a 1999 consensus. Arch Environ Health 54: 147-149

Bascom R (1992) Multiple chemical sensitivity: a respiratory disorder? Toxicol Ind Health 8: 221-8

Bauer A, Schwarz E, Mai C, Hauf O (2006) Long time follow-up of patients with environmental illness or multiple chemical sensitivity (MCS). Proceedings of the Meeting of the German Society for Epidemiology, Greifswald, Germany. 21-23 September 2006: 272

Bauer A, Schwarz E and Mai C (2008) Multiple Chemical Sensitivity/MCS: Ein Update. Umwelt Medizin Gesellschaft 21: 9-15

Bell IR, Baldwin CM, Fernandez M, Schwartz GE (1999a) Neural sensitization model for multiple chemical sensitivity: overview of theory and empirical evidence. Toxicol Ind Health 15: 295-304.

Bell IR, Baldwin CM, Schwartz GE (2001) Sensitization studies in chemically intolerant individuals: implications for individual difference research. Ann N Y Academy of Science 933: 38-47.

Bell IR, Schwartz GE, Peterson JM and Amend D (1993b) Symptom and personality profiles of young adults from a college student population with self-reported illness from foods and chemicals. J Am Coll Nutr 12: 693-702

Bell IR, Schwartz GE, Peterson JM, Amend D and Stini W (1993c) Possible time-dependent sensitization to xenobiotics: self-reported illness from chemical odours, foods, and opiate drugs in an older adult population. Arch Environ Health 48: 315-327

Bell IR, Szarek MJ, Dicenso DR, Baldwin CM, Schwartz GE, Bootzin RR (1999b) Patterns of waking EEG spectral power in chemically intolerant individuals during repeated chemical exposures. Int J Neurosci. 97: 41-59

Bell IR, Warg-Damiani L, Baldwin CM, Walsh ME and Schwartz GE (1998) Self-reported chemical sensitivity and wartime chemical exposures in Gulf War veterans with and without decreased global health ratings. Military Med 163: 725-732

Berg ND, Rasmussen HB, Linneberg A, Brasch-Andersen C, Fenger M, Dirksen A, Verterhauge S, Werge T and Elberling J (2010) Genetic susceptibility factors for multiple chemical sensitivity revisited. Int J Hyg Environ Health 213: 131-139

Berkowitz GS, et al. (2003) Exposure to indoor pesticides during pregnancy in a multiethnic, urban cohort. Environ Health Perspect 111:79–84.

Berkowitz GS, et al. (2004) In utero pesticide exposure, maternal paraoxonase activity, and head circumference. Environ Health Perspect 112:388–391.

Bessac BF and Jordt S-E (2008) Breathtaking TRP channels: TRPA1 and TRPV1 in airway chemosensation and reflex control. Physiology (Bethesda) 23: 360-370

Black DW, Rathe A and Goldstein RB (1990) Environmental illness: A controlled study of 26 patients with 20th century disease. JAMA 264: 3166-3170

Black DW, Doebbeling BN, Voelker MD, Clark WR, Woolson RF, Barrett DH and Schwartz DA (1999) Quality of life and health-services utilization in population-based sample of military personnel reporting multiple chemical sensitivity. J. Occup. Environ. Med. 41: 928-33

Black DW, Doebbeling BN, Voelker MD, Clarke WR, Woolson RF, Barrett DH and Schwartz DA (2000a) Multiple chemical sensitivity syndrome: symptom prevalence and risk factors in a military population. Arch Intern Med 160: 1169-76

Black DW, Okiishi C and Schlosser S (2000c) The Iowa follow-up of chemically sensitive persons. Annals NY Acad Sci 933:48-56

Board of Environmental Studies and Toxicology, Commission on Life Sciences, National Research Council (1992) Multiple Chemical Sensitivities. A Workshop. National Academy Press, Washington DC, 1992.

Bock KW and Birbaumer N (1997) MCS (multiple chemical sensitivity): Cooperation between toxicology and psychology may facilitate solutions of the problems: commentary. Human Exp Toxicol 16: 481-484

Bornschein S, Hausteiner C, Drzezga A, Bartenstein P, Schwaiger M, Forstl H and Zilker T (2002b) PET in patients with clear-cut multiple chemical sensitivity (MCS). Nuklearmedizin 41(6): 233-9

Bornschein S, Hausteiner C, Drzezga A, Theml T, Heldmann B, Grimmer T, Perneczky R, Jahn T, Schwaiger M, Zilker T, Förstl H (2007) Neuropsychological and positron emission tomography correlates in idiopathic environmental intolerances. Scand J Work Environ Health. 33: 447-53.

Bornschein S, Hausteiner C, Pohl C, Jahn T, Angerer J, Foerstl H and Zilker T (2008) Pest controllers: A high risk group for multiple chemical sensitivity (MCS)? Clin. Toxicol. 46:

193-200

BouchardMF, et al. (2011) Prenatal exposure to organophosphate pesticides and IQ in 7-year-old children. Environ Health Perspect 119:1189–1195.

Bradman A, et al. (2003) Measurement of pesticides and other toxicants in amniotic fluid as a potential biomarker of prenatal exposure: A validation study. Environ Health Perspect 111:1779–1782.

Bradway DE, Shafik TM, Lores EM. (1997) Comparison of cholinesterase activity, residue levels, and urinary metabolite excretion of rats exposed to organophosphorus pesticides. J Agric Food Chem. 1977;25(6):1353–1358CrossRefMedlineWeb of Science29

Bravo R, Caltabiano LM, Weerasekera G, et al. (2004) Measurement of dialkyl phosphate metabolites of organophosphorus pesticides in human urine using lyophilization with gas chromatography-tandem mass spectrometry and isotope dilution quantification. J Expo Anal Environ Epidemiol. 2004;14(3):249–259CrossRefMedlineWeb of Science25

Brown-DeGagne AM and McGlone J (1999) Multiple chemical sensitivity: a test of the olfactory-limbic model. J. Occup. Environ. Med. 41: 366-77 - 86 - A Review of Multiple Chemical Sensitivity

Buchwald D and Garrity D (1994) Comparison of patients with chronic fatigue syndrome, fibromyalgia, and multiple chemical sensitivities. Arch Intern Med 154: 2049-2053

Buist, R. Food and Chemical Sensitivity. Harper & Row, Sydney, 1986. Cannon, Geoffrey

Burge PS (2004) Sick Building Syndrome. Occup Environ Med. 61: 185-190

Caccappolo E, Kipen H, Kelly-McNeil K, Knasko S,Hamer RM, Natelson B and Fiedler N (2000) Odour perception: multiple chemical sensitivities, chronic fatigue, and asthma. J Occup Environ Med. 42: 629-38

California Medical Association Scientific Board Task Force on Clinical Ecology (1986) Clinical ecology: a critical appraisal. West J Med 144: 239-245

Camps J, Marsillach J and Joven J (2009) The paraoxonases: role in human diseases and methodological difficulties in measurement. Crit Rev Clin Lab Sci 46: 83-106

Canadian Centre for Occupational Health and Safety (2008) Indoor Air Quality: A Legitimate OSH Concern. http://www.ccohs.ca/headlines/text27.html. Accessed October 2009.

Caress SM and Steinemann AC (2003) A review of a two-phase population study of multiple chemical sensitivities. Environ Health Perspec 111: 1490-1497

Caress SM and Steinemann AC (2004) Prevalence of multiple chemical sensitivities: A Population-based study in the South-eastern United States. Am. J. Pub. Health. 94: 746-747

Chaboussou, F., 1985 (English 2004 trans.), Healthy Plants, A New Agricultural Revolution, Jon Carpenter, Charlbury, UK.

Chanda SM, Pope CN. Neurochemical and neurobehavioral effects of repeated gestational exposure to chlorpyrifos in maternal and developing rats. Pharmacol Biochem Behav. 1996 Apr;53(4):771–776

Chiodo LM, Jacobson SW, Jacobson JL (2004) Neurodevelopmental effects of postnatal lead exposure at very low levels. Neurotoxicol Teratol 26:359–371.

Cohen Hubal EA, Sheldon LS, Burke JM, et al. (2000) Children's exposure assessment: a review of factors influencing Children's exposure, and the data available to characterize and assess that exposure. Environ Health Perspect. 2000;108(6):475–486Medline3.

Creaser W, Miller P, Hogan A, Kyaw-Myint S, Hill J, May B and Stavreski B (2007) The Australian hazard exposure assessment database. J Occup Health Safety 23: 563-570

Cullen MR, Pace PE and Redlich CA (1992) The experience of the Yale occupational and environmental medicine clinics with multiple chemical sensitivities, 1986-1991. Toxicol Ind Health 8(4): 15-19

Cullen MR (1987) Workers with multiple chemical sensitivities. Occup Med: State of the Art Reviews 2: 655-661

Dalton P and Hummel T (2000) Chemosensory function and response in idiopathic environmental intolerance. Occup Med: State of the Art Reviews 15: 539-556

Das-Munshi J, James Rubin G and Wessely S (2006) Multiple chemical sensitivities: A systematic review of provocation studies. J. Allergy Clin Immunol 118: 1257-1264 - 87 - A Review of Multiple Chemical Sensitivity

Das-Munshi J, James Rubin G and Wessely S (2007) Multiple chemical sensitivity: review. Current Opinion in Otolaryngology and Head and Neck Surgery 15: 274-280

De Luca C, Scordo MG, Cesareo E, Pastore S, Mariani S, Maiani G, Stancato A, Loreti B,

Valacchi G, Lubrano C, Raskovic D, De Padova L, Genovesi G and Korkina LG (2010) Biological definition of multiple chemical sensitivity from redox state and cytokine profiling and not from polymorphisms of xenobiotic-metabolizing enzymes. Toxicol Appl Pharmacol (In press).

Donnay AH (1998) Recognition of multiple chemical sensitivity. MCS Referral and Resources. http://www.mcsrr.org/factsheets/MCSrecogn.pdf. Accessed October 2009.

Donnay AH (1999) On the recognition of multiple chemical sensitivity in medical literature and government policy. Int J Toxicol 18: 383-392

Donnay A and Ziem G (1995) Protocol for evaluating disorders of porphyrin metabolism in Chemical Sensitive Patients, MCS referral and Resources, Baltimore.

Donoghue AM and Cullen MR (2007) Air emissions from Wagerup alumina refinery and community symptoms: An environmental case study. Journal of Occupational and Environmental Medicine 49: 1027-1039 1

Doty RL, Deems DA, Frye RE, Pelberg R and Shapiro A (1988) Olfactory sensitivity, nasal resistance, and autonomic function in patients with multiple chemical sensitivities. Arch Otolaryngol Head Neck Surg 114: 1422-1427

Doty RL (1994) Olfaction and multiple chemical sensitivity. Toxicol Ind Health 10: 359-368

Driscoll T, Mannetje A, Dryson E, Feyer A-M, Gander P, McCracken S, Pearce N, Wagstaffe M (2004) The burden of occupational disease and injury in New Zealand: Technical Report. NOHSAC: Wellington.

Eaton KK, Anthony HM, Birtwistle S, Downing D, Freed DLJ, McLaren Howard J, Maberly DJ, Mansfield JR, Myhill S and Radcliffe MJ (2000) Multiple chemical sensitivity: recognition and management. A document on the health effects of everyday chemical exposures and their implications. J Nutr Environ Med 10: 39-84

Eberlein-Konig B, Przybilla B, Kuhnl P, Golling G, Gebefugi I and Ring J (2002) Multiple chemical sensitivity (MCS) and others: Allergological, environmental and psychological investigations in individuals with indoor air related complaints. Int J Hyg Environ Health 205: 213-220

Eis D, Helm D, Mühlinghaus T, Birkner N, Dietel A, Eikmann T, Gieler U, Herr C, Lacour M, Nowak D, Pedrosa Gil F, Podoll K, Renner B, Andreas Wiesmüller G, Worm M (2008) The German Multicentre Study on Multiple Chemical Sensitivity (MCS). Int J Hyg Environ Health 211: 658-81

Engel SM, et al. (2007) Prenatal organo-phosphate metabolite and organochlorine levels and performance on the Brazelton Neonatal Behavioral Assessment Scale in a multiethnic pregnancy cohort. Am J Epidemiol 165:1397–1404.

Eskenazi B, Marks AR, Bradman A, et al. (2007) Organophosphate pesticide exposure and neurodevelopment in young Mexican-American children. Environ Health Perspect. 2007;115(5):792–798MedlineWeb of Science13.

Faustman EM, Silbernagel SM, Fenske RA, Burbacher TM, Ponce RA. (2000) Mechanisms underlying Children's susceptibility to environmental toxicants. Environ Health Perspect. 2000 Mar;108 (Suppl 1):13–21

Fernandez M, Bell IR and Schwartz GER (1999) EEG sensitisation during chemical exposure in women with and without chemical sensitivity of unknown etiology. Toxicol Ind Health 15:305-312

Fiedler N and Kipen HM (1997) Chemical sensitivity: the scientific literature. Environ Health Perspect 105(Suppl 2): 409-415

Fiedler N and Kipen HM (2001) Controlled exposures to volatile organic compounds in sensitive groups. Ann NY Acad. Sci 933: 24-37

Fiedler N, Maccia C and Kipen H (1992) Evaluation of chemically sensitive patients. J Occup Med 34: 529-538

Fitzgerald DJ (2008) Studies on self-reported multiple chemical sensitivity in South Australia. Environmental Health 8: 33-39

Friedman A, Kaufer D, Shemer J, Hendler I, Soreq H and Tur-Kaspa I (1996) Pyridostigmine brain penetration under stress enhances neuronal excitability and induces early immediate transcriptional response. Nat Med 2: 1382-1385

Fukuyama T, Ueda H, Hayashi K, Tajima Y, Shuto Y, Saito TR, Harada T, Kosaka T (2008) Detection of low-level environmental chemical allergy by a long-term sensitization method. Toxicol Lett 180: 1-8.

Gad SC (1999) Multiple chemical sensitivity: a moderator's viewpoint. Int J Toxicol 18: 379-381

Garcia SJ, Seidler FJ, Crumpton TL, Slotkin TA (2001) Does the developmental neurotoxicity of chlorpyrifos involve glial targets? Macromolecule synthesis, adenylyl cyclase signaling, nuclear transcription factors, and formation of reactive oxygen in C6 glioma cells. Brain Res 891:54–68.

Georgellis A, Lindelof B, Lundin A, Arnetz B and Hillert L (2003). Multiple chemical sensitivity in male painters; a controlled provocation study. Int J Hyg Environ Health 206: 531-538

Gibson PR, Elms AN, Ruding LA (2003) Perceived treatment efficacy for conventional and alternative therapies reported by persons with multiple chemical sensitivity. Environ Health Perspect 111: 1498-1504.

Gilbert ME (1995) Repeated exposure to lindane leads to behavioral sensitization and facilitates electrical kindling. Neurotoxicol Teratatol 17: 131-141

Grandjean P, Harari R, Barr DB, Debes F. (2006)Pesticide exposure and stunting as independent predictors of neurobehavioral deficits in Ecuadorian school children. Pediatrics. 2006;117(3).

Graveling RA, Pilkington A, George JPK, Butler MP and Tannahill SN (1999) A review of multiple chemical sensitivity. Occup Environ Med 56: 73-85

Gray GC, Reed RJ, Kaiser KS, Smith TC and Gastaflaga VM (2002) Self-reported symptoms and medical conditions among 11,868 Gulf War-era veterans. The Seebee health study. Am J Epidemiol 155: 1033-1044

Groneberg DA, Niimi A, Thai Dinh Q, Cosio B, Hew M, Fischer A and Fan Chung K (2004)

Increased expression of transient receptor potential vanilloid-1 in airway nerves of chronic cough. Am J Resp Crit Care Med 170: 1276-1280

Haley RW, Billecke S and La Du BN (1999) Association of low PON1 type Q (type A) arylesterase activity with neurologic symptom complexes in Gulf War veterans. Toxicol.Appl Pharmacol 157: 227-233

Heuser G, Mena I and Alamos F (1994) Neurospect findings in patients exposed to neurotoxic chemicals. Toxicol Ind Health 10: 561- 572

Heuser G, Wojdani A and Heuser S (1992) Diagnostic markers of multiple chemical sensitivity. Multiple chemical sensitivities: addendum to biologic markers in immunotoxicology. Washington, DC: National Academy Press.

Heuser G, Wu JC (2001) Deep subcortical (including limbic) hypermetabolism in patients with chemical intolerance: human PET studies. Ann N Y Acad Sci. 933: 319-22.

Hillert L, Musabasic V, Berglund H, Ciumas C and Savic I (2007) Odor processing in multiple chemical sensitivity. Human Brain Mapping 28: 172-182

Hodgson M (2000) Sick building syndrome. Occup Med 15: 571-85

Hojo S, Ishikawa S, Kumano H, Miyata M and Sakabe K (2008) Clinical characteristics of physician-diagnosed patients with multiple chemical sensitivity in Japan Int J Environ Health 211: 682-689

Holmes GP, Kaplan JE, Gantz NM, Komaroff AL, Schonberger LB, Straus SE, Jones JF, Dubois RE, Cunningham-Rundles C, Pahwa S, et al. (1988) Chronic fatigue syndrome: a working case definition. Ann Intern Med 108: 387-389

Hummel T, Roscher S, Jaumann JP and Kobal G (1996) Intranasal chemoreception in patients with multiple chemical sensitivities: a double-blind investigation. Regul Toxicol Pharmacol 24: S79-S86

Jason L, Taylor RR and Kennedy CL (2000) Chronic fatigue syndrome, fibromyalgia, and multiple chemical sensitivities in a community-based sample of persons with chronic fatigue syndrome-like symptoms. Psychosom Med 62: 655-663

Jurewicz J, Hanke W. (2008) Prenatal and childhood exposure to pesticides and neurobehavioral development: review of epidemiological studies. Int J Occup Med Environ Health. 2008;21(2):121–132CrossRefMedlineWeb of Science11

Joffres MR, Sampalli T and Fox RA (2005) Physiologic and symptomatic responses to low level substances in individuals with and without chemical sensitivities: A randomized controlled blinded pilot booth study. Environ Health Perspect 113: 1178-1183

Joffres MR, Williams T, Sabo B and Fox RA (2001) Environmental sensitivities: Prevalence of major symptoms in a referral centre: The Nova Scotia environmental sensitivities research centre study. Environ Health Perspect 109: 161-165

Kassirer J, Koswan S, Spence K, Morphet S, Wolnik, C, Goom S and Del Matto T (2004) The Impact of By-Laws and Public Education Programs on Reducing the Cosmetic / NonEssential, Residential Use of Pesticides: A Best Practices Review. Prepared by CULLBRIDGE™ Marketing and Communications and Canadian Centre for Pollution Prevention.

Kimata H (2004) Effect of exposure to volatile organic compounds on plasma levels of neuropeptides, nerve growth factor and histamine in patients with self-reported multiple chemical sensitivity. Int. J. Environ. Health 207: 159-163

Kipen HM and Fiedler N (2002) The role of environmental factors in medically unexplained symptoms and related syndromes: The evidence and the challenge. Environ Health Perspect 110: 597-599

Kipen H, Fiedler N, Maccia C, Yurkow E, Todaro J and Laskin D (1992) Immunological evaluation of chemically sensitive patients. Toxicology and Industrial Health 8: 125-135

Kipen HM, Hallman W, Kang H, Fiedler N and Natelson BH (1999) Prevalence of chronic fatigue and chemical sensitivities in Gulf registry veterans. Arch Environ Health 54: 313-318

Kipen HM, Hallman W, Kelly-McNeil K and Fiedler N (1995) Measuring chemical sensitivity prevalence: a questionnaire for population studies. Am J Public Health 85: 574-577

Kovacic P. (2003) Mechanism of organo-phosphates (nerve gases and pesticides) and antidotes: electron transfer and oxidative stress. Curr Med Chem 2003; 10: 2705-2709

Kreutzer R (2002) MCS. The status of population-based research. Int J Hyg Envir Health

205: 411-414 Kreutzer R, Neutra RR and Lashuay N (1999) Prevalence of people reporting sensitivities to chemicals in a population. Am J Epidemiol 150: 1-12

Kreutzer R (2000) Idiopathic environmental intolerance: case definition issues. Occup Med 15(3): 511-7 - 92 - A Review of Multiple Chemical Sensitivity

Labarge XS and McCaffrey RJ (2000) Multiple chemical sensitivity: a review of the theoretical and research literature. Neuropsychol Rev 10(4):183-211.

Lacour M, Zunder T, Schmidtke K, Vaith P and Scheidt C (2005) Multiple chemical sensitivity syndrome (MCS) – suggestions for an extension of the US MCS-case definition. Int. J. Hyg. Environ.-Health 208: 141-151

Landrigan PJ, Claudio L, Markowitz SB, Berkowitz GS, Brenner BL, Romero H, Wetmur JG, Matte TD, Gore AC, Godbold JH, et al. (1999) Pesticides and inner-city children: exposures, risks, and prevention. Environ Health Perspect. 1999 Jun;107 (Suppl 3):431–437

Lawler R. (1991) Voices of the First Day. Inner Traditions International.

Lax MB and Henneberger PK (1995) Patients with multiple chemical sensitivities in an occupational health clinic: presentation and follow-up. Arch Environ Health 50: 425-431

Lehrer PM (1997) Psychophysiological hypotheses regarding multiple chemical sensitivity syndrome. Environ Health Perspect 105: 479-483

Lehrer PM (2000) Behaviour conditioning and idiopathic environmental intolerance. Occup Med 15(3): 519-528

Lessof M (1997) Meeting Report. Report of multiple chemical sensitivities (MCS) workshop, Berlin, Germany, 21-23 February

1996. PCS/96.29 IPCS, Geneva, Switzerland. Human and Experimental Toxicology 16: 233-234

Levallois P (2002) Hypersensitivity of human subjects to environmental electric and magnetic field exposure: A review of the literature. Environ. Health Perspect. 110 (Suppl4): 613-618

Levin AS and Byers VS (1987) Environmental illness: a disorder of immune regulation. Occup Med 2(4): 669-681

Levin AS and Byers VS (1992) Multiple chemical sensitivities: a practicing clinician's point of view- clinical and immunologic research findings. Proceedings of the AOEC Workshop on Multiple Chemical Sensitivity. Toxicol Ind Health 8(4): 95-109

Levy F (1997) Clinical features of multiple chemical sensitivity. Scand. J. Work Environ. Health 23 (suppl 3): 69-73

Leznoff A and Binkley KE (2000) Idiopathic environmental intolerances: results of challenge studies. Occup Med 15(3): 529-37

Leznoff A (1997) Clinical aspects of allergic disease: Provocation challenges in patients with multiple chemical sensitivity. J Allergy Clin Immunol 99(4): 438-442

Lorig TS (1994) EEG and ERO studies of low level odour exposure in normal subjects. Toxicol Industrial Health 10: 579-86

Malo J-L, L'Archevêque J, Castellanos L, Lavoie K, Ghezzo H and Maghni K (2009) Longterm outcomes of acute irritant-induced asthma. Am J Respir Crit Care Med 179: 923-928

Mauro RE, Zhang L (2007) Unique insights into the actions of CNS agents: Lessons from studies of chlorpyrifos and other common pesticides. Central Nerv Syst Agents Med Chem 7(3):183–199

Mayberg H (1994) Critique: SPECT studies of multiple chemical sensitivity. Toxicol Ind Health 10(4/5): 661-665

McKelvey A, Ashe R, McKenna D, Roberts R. (2010) Caesarean section in the second stage of labour: a retrospective review of obstetric setting and morbidity. J Obstet Gynaecol. 2010 Apr;30(3):264-7.

McKeown-Eyssen GE, Baines CJ, Cole DEC, Riley N, Tyndale RF, Marshall LM and Jazmaji V (2004) Case-control study of genotypes in multiple chemical sensitivity; CYP2D6, NAT1, NAT2, PON1, PON2 and MTHFR. Int J Epidemiol 33: 1-8

McKeown-Eyssen GE, Baines CJ, Marshall LM, Jazmaji V and Sokoloff ER (2001) Multiple chemical sensitivity: discriminant validity of case definitions. Arch Environ Health 56(5):406-12

Meggs WJ and Cleveland CH Jr (1993) Rhinolaryngoscopic examination of patients with the multiple chemical sensitivity syndrome. Arch Environ Health 48(1): 14-18

Meggs WJ, Dunn KA, Bloch RM, Goodman PE and Davidoff AL (1996) Prevalence and nature of allergy and chemical sensitivity in a general population. Arch Environ Health 51(4): 275–82

Meggs WJ (1993) Neurogenic inflammation and sensitivity to environmental chemicals. Environ Health Perspect 101: 234-8

Meggs WJ (1995) Neurogenic switching: a hypothesis for a mechanism for shifting the site of inflammation in allergy and chemical sensitivity. Environ Health Perspective 103: 54-6

Meggs WJ (1999) Mechanisms of allergy and chemical sensitivity. Toxicol Ind Health 15: 331-338

Meggs WJ (1992) MCS and the immune system. Toxicol Ind Health 8(4): 203-214

Miller C, Ashford N, Doty R, Lamielle M, Otto D, Rahill A and Wallace L (1997) Empirical approaches for the investigation of toxicant-induced loss of tolerance. Environ Health Perspect 102 (Suppl 2): 515-519

Miller CS and Mitzel HC (1995) Chemical sensitivity attributed to pesticide exposure versus remodeling. Arch Environ Health 50: 119-129

Miller CS and Prihoda TJ (1999b) The Environmental Exposure and Sensitivity Inventory (EESI): a standardized approach for measuring chemical intolerances for research and clinical applications. Toxicol Ind Health 15:370-385

Miller CS (1992) Possible models for multiple chemical sensitivity: conceptual issues and the role of the limbic system. Toxicol Ind Health 8: 181-190

Miller CS (1996) Chemical sensitivity: symptom, syndrome or mechanism for disease? Toxicol 111: 69-86

Miller CS (1997) Toxicant-induced loss of tolerance-An emerging theory of disease? Environ Health Perspect 105 (Suppl 2): 445-453

Miller CS (2000) Toxicant- induced loss of tolerance. Addiction 96: 115-139

Miller CS and Mitzel HC (1995) Chemical sensitivity attributed to pesticide exposure versus remodelling. Arch Environ Health 50: 119-128

Miller CS and Prihoda TJ (1999a) A controlled comparison of symptoms and chemical intolerances reported by Gulf War veterans, implant recipients and persons with multiple chemical sensitivity. Toxicol Ind Health 15: 386-397

Millqvist E (2008) Mechanisms of increased airway sensitivity to occupational chemicals and odors. Curr Opinion Allergy Clin Immunol 8: 135-139

Mitchell CS, Donnay A, Hoover DR and Margolick JB (2000) Immunologic parameters of multiple chemical sensitivity. Occup Med: State of the Air Reviews 15(3): 647-65

Mitchell FL (1995) Multiple Chemical Sensitivity: A Scientific Overview. Frank Mitchell (ed) , Princeton Scientific Publishing Co.

Mooser SB (1987) The epidemiology of multiple chemical sensitivities. Occup Med: State of the Art Reviews 2: 663-668

Müller KE and Schnakenberg E (2008) Die bedeutung der glukuronidierung bei umweltmedizinischen erkrankungen am beispiel der UDP-glukurinosyltransferase 1A1. Umwelt-Medizin-Gesellschaft 21: 295-300

Nassini R, Materazzi S, De Siena G, De Cesaris F and Geppetti P (2010) Transient receptor potential channels as novel drug targets in respiratory diseases. Curr Opin Investig Drugs 11: 535-542

Nethercott JR, Davidoff LL, Curbow B and Abbey H (1993) Multiple chemical sensitivities syndrome: toward a working case definition. Arch Environ Health 48: 19-26

Nielsen GD (1991) Mechanisms of activation of the sensory irritant receptor by airborne chemicals. Crit Rev Toxicol 21: 183-208

Ojima M, Tonori H, Sato T, Sakabe K, Miyata M, Ishikawa S and Y Aizawa (2002) Odour perception in patients with multiple chemical sensitivity. Tohoku J Exp Med 198: 163-173

Orme T and Benedetti P (1994) Multiple Chemical Sensitivity. Prepared for the American Council on Science and Health. http://www.acsh.org/publications/pubID.847/pub_detail.asp

Orriols R, Costa R, Cuberas G, Jacas C, Castell J and Sunyer J (2009) Brain dysfunction in multiple chemical sensitivity. J. Neurol. Sci. 287: 72-78

Österberg K, Orbaek P, Karlson B, Akesson B and Bergendorf U (2003) Annoyance and performance during the experimental chemical challenge of subjects with multiple chemical sensitivity. Scand J Work Environ Health 29(1): 40-50

Overstreet DH and Djuric V (2001) A genetic rat model of cholinergic hypersensitivity: implications for chemical intolerance, chronic fatigue, and asthma. Ann N Y Acad Sci. 933: 92-102

Pall ML (2001) Common etiology of posttraumatic stress disorder, fibromyalgia, chronic fatigue syndrome and multiple chemical sensitivity via elevated nitric oxide/peroxynitrite. Med Hypotheses 57: 139-45

Pall ML (2002) NMDA sensitisation and stimulation by peroxynitrite, nitric oxide, and organic solvents as the mechanism of chemical sensitivity in MCS. FASEB 16:1407-1417

Pall ML (2003) Elevated nitric oxide/peroxynitrite theory of multiple chemical sensitivity: Central role of N-methyl-D-aspartate receptors in the sensitivity mechanism. Environ Health Perspec 111(12): 1461-4

Pall ML (2004) The vanilloid receptor as a putative target of diverse chemicals in multiple chemical sensitivity. Arch. Environ. Health 59: 363-375

Pall ML (2006) The NO/ONOO-cycle as the Cause of Fibromyalgia and Related Illnesses: Etiology, Explanation and Effective Therapy. In: New Research on Fibromyalgia. JA

Pederson (Ed.), Nova Science Publishers, New York. pp39-59

Pall ML (2007b) Explaining "Unexplained Illnesses". Disease Paradigm for Chronic Fatigue Syndrome, Multiple Chemical Sensitivity, Fibromyalgia, Post-Traumatic Stress Disorder, Gulf War Syndrome, and Others. Harrington Park Press, New York.

Pall ML (2009) Multiple Chemical Sensitivity: Toxicological Questions and Mechanisms. In: Bryan Ballantyne, Timothy C. Marrs, Tore Syversen (Eds), General and Applied Toxicology, 3rd Edition, John Wiley and Sons.

Park J and Knudson S (2007) Medically unexplained physical symptoms. Health Reports 18:43-47. Statistics Canada, Catalogue 82-003

Perera FP, et al. (2002) The challenge of preventing environmentally related disease in young children: Community-based research in New York City. Environ Health Perspect 110:197–204.

Perera FP, et al. (2003) Effects of transplacental exposure to environmental pollutants on birth outcomes in a multiethnic population. Environ Health Perspect 111:201–205.

Plessen KJ, et al. (2006) Hippocampus and amygdala morphology in attention-deficit/hyperactivity disorder. Arch Gen Psychiatry 63:795–807.

Randolph TG (1961) Human ecology and susceptibility to the chemical environment. Annals Allergy 19: 518-540

Rauh VA, et al. (2006) Impact of prenatal chlorpyrifos exposure on neurodevelopment in the first 3 years of life among inner-city children. Pediatrics 118(6):e1845–e1859.

Rauh V, et al. (2011) Seven-year neurodevelopmental scores and prenatal exposure to chlorpyrifos, a common agricultural pesticide. Environ Health Perspect 119:1196–1201.

Rauh VA, Perera FP, Horton MK, Whyatt RM, Bansal R, Hao X, Liu J, Barr DB, Slotkin TA, Peterson BS. (2012) Brain anomalies in children exposed prenatally to a common organo-phosphate pesticide. Proc Natl Acad Sci U S A. 2012 May 15; 109(20):7871-6.

Rea WJ, Johnson AR, Ross GH, Butler JR, Fenyves EJ, Griffiths B and Laseter J (1992) Considerations for the diagnosis of chemical sensitivity. In: Multiple Chemical Sensitivities. Washington DC, National Academy Press.

Read D (2002) Multiple Chemical Sensitivities. Report for Environmental Risk Management Authority (ERMA), New Zealand.

Reid S, Hotopf M, Hull L, Ismail K, Unwin C and Wessely S (2001) Multiple chemical sensitivity and chronic fatigue syndrome in British gulf war veterans. Am J Epidemiol 153(6): 604-9

Rest K (1992) Advancing the understanding of multiple chemical sensitivity (MCS). Overview and recommendations from an AOEC workshop. Toxicology and Industrial Health 8: 1-13

Rohlman DS, Arcury TA, Quandt SA, et al. (2005) Neurobehavioral performance in preschool children from agricultural and non-agricultural communities in Oregon and North Carolina. Neurotoxicology. 2005;26(4):589–598CrossRefMedlineWeb of Science16.

Ross GH (1992) History and clinical presentation of the chemically sensitive patient. Toxicology and Industrial Health 8: 21-28

Ross GH, Rea WJ, Johnson AR, Hickey DC and Simon TR (1999) Neurotoxicity in Single Photon Emission Computed Tomography brain scans of patients reporting chemical sensitivity. Toxicol Ind Health 15: 415-420

Roy TS, Seidler FJ, Slotkin TA (2004) Morphologic effects of subtoxic neonatal chlorpyrifos exposure in developing rat brain: regionally selective alterations in neurons and glia. Brain Res Dev Brain Res 148:197–206.

Roy TS, Sharma V, Seidler FJ, Slotkin TA (2005) Quantitative morphological assessment reveals neuronal and glial deficits in hippocampus after a brief subtoxic exposure to chlorpyrifos in neonatal rats. Brain Res Dev Brain Res 155:71–80.

Royal College of Physicians and Royal College of Pathologists (1995) Good allergy practice -standards of care for providers and purchasers of allergy services within the National Health service. Clin Exp Allergy 25: 586-595

Ruckart PZ, Kakolewski K, Bove FJ, Kaye WE. (2004) Long-term neurobehavioral health effects of methyl parathion exposure in children in Mississippi and Ohio. Environ Health Perspect. 2004;112(1):46–51MedlineWeb of Science15

Schnakenberg E, Fabig K-R, Stannula M, Strobl N, Lustig M, Fabig N and Schloot W (2007) A cross-sectional study of self-reported chemical-related sensitivity is associated with gene variants of drug-metabolizing enzymes. Environmental Health 6: 6-16

Simon TR, Hickey DC, Fincher CE, Johnson AR, Ross GH and Rea WJ (1994) Single proton emission computed tomography of the brain in patients with chemical sensitivities. Toxicol Ind Health 10: 573-577

Shahin Shadnia, Mojgan Dasgar, et al. (2007) Protective Effects of α-Tocopherol and N-Acetyl-Cysteine on Diazinon- Induced Oxidative Stress and Acetylcholinesterase Inhibition in Rats Toxicology Mechanisms and Methods 2007 17:2, 109-115

Slotkin TA (2004) Guidelines for developmental neurotoxicity and their impact on organo-phosphate pesticides: A personal view from an academic perspective. Neurotoxicology 25:631-640.

Social Development Committee (2005) Inquiry into Multiple Chemical Sensitivity. Twenty Second Report of the Social Development Committee. Parliament of South Australia

Slotkin TA (2005) Developmental neurotoxicity of organo-phosphates: A case study of chlorpyrifos. Toxicity of Organo-phosphate and Carbamate Pesticides, ed Gupta RC (Elsevier Academic, San Diego), pp 293–314.

Slotkin TA, Seidler FJ. (2007) Comparative developmental neurotoxicity of organo-phosphates in vivo: transcriptional responses of pathways for brain cell development, cell signaling, cytotoxicity and neurotransmitter systems. Brain Res Bull. 2007;72(4–6):232–274CrossRefMedlineWeb of Science34.

Slotkin TA, Seidler FJ (2005) The alterations in CNS serotonergic mechanisms caused by neonatal chlorpyrifos exposure are permanent. Brain Res Dev Brain Res 158:115–119.

Slotkin TA. (2004) Cholinergic systems in brain development and disruption by neurotoxicants: nicotine, environmental tobacco smoke, organo-phosphates. Toxicol Appl Pharmacol. 2004;198(2):132–151CrossRefMedlineWeb of Science

Smith CV, Jones DP, Guenthner TM, Lash LH, Lauterburg BH. (1996) Department of Pediatrics, Baylor College of Medicine, Houston 77030, USA. Toxicology and Applied Pharmacology, 1996, 140(1):1-12

Smith V. (1982) Anaphylactic Shock, Acute Renal Failure, and Disseminated Intravascular Coagulation: Suspected Complications of Zomepirac. JAMA. 1982;247(8):1172-1173

Solomon G and Moodley J. (2007) Acute chlorpyrifos poisoning in pregnancy: A case report Clinical Toxicology 45:4, 416-419

Sorg BA (1999) Multiple chemical sensitivity: potential role for neural sensitization. Crit Rev Neurobiol 13(3): 283316

Sorg BA, Tschirgi ML, Swindell S, Chen L and Fang J (2001) Repeated formaldehyde effects in an animal model for multiple chemical sensitivity. Ann. NY Acad. Sci. 933: 57-67

Sparks PJ, Daniell W, Black DW, Kipen HM, Altman LC, Simon GE and Terr AI (1994)Multiple chemical sensitivity syndrome: a clinical perspective. 1. Case definition, theories of pathogenesis, and research needs. J. Occup. Med. 36: 718-730

Sparks PJ (2000a) Diagnostic evaluation and treatment of the patient presenting with idiopathic environmental intolerance. Occup Med 15: 601-9

Sparks PJ (2000b) Idiopathic environmental intolerances: overview. Occup Med 15(3): 497-510

Sultatos LG. (1994) Mammalian toxicology of organophosphorus pesticides. J Toxicol Environ Health. 1994;43(3):271–289MedlineWeb of Science31

Tafuri J, Roberts J. (1987) Organo-phosphate Poisoning, Annals of Emergency Medicine. Volume 16, Issue 2 , Pages 193-202, February 1987

Tai C, Zhu S and Zhou N (2008) TRPA1: The central molecule for chemical sensing in pain pathway? J Neurosci 28: 1019-1021

Timofeeva OA, Roegge CS, Seidler FJ, Slotkin TA, Levin ED. (2008) Persistent cognitive alterations in rats after early postnatal exposure to low doses of the organo-phosphate pesticide, diazinon. Neurotoxicol Teratol. 2008;30(1):38–45CrossRefMedlineWeb of Science36.

Timofeeva OA, Sanders D, Seemann K, et al. (2008) Persistent behavioral alterations in rats neonatally exposed to low doses of the organo-phosphate pesticide, parathion. Brain Res Bull. 2008;77(6):404–411CrossRefMedlineWeb of Science35

Thomas HV, Stimpson NJ, Weightman AL, Dunstan F and Lewis G (2006) Systematic review of multi-symptom conditions in Gulf War veterans. Psychological Med. 36: 735-747

Thrasher JD, Broughton A and Madison R (1990) Immune activation and autoantibodies in humans with long-term inhalation exposure to formaldehyde. Arch Environ Health 45: 217-223

Waddell W (1993) The science of toxicology and its relevance to MCS. Regul Tox Pharmacol 18: 13-22

Wadia RS. Treatment of Organo-phosphate Poisoning. Indian J Crit Care Med 2003;7:85-7

Whyatt RM, et al. (2002) Residential pesticide use during pregnancy among a cohort of urban minority women. Environ Health Perspect 110:507–514.

Whyatt RM, et al. (2004) Prenatal insecticide exposures and birth weight and length among an urban minority cohort. Environ Health Perspect 112:1125–1132.

Whyatt RM, et al. (2005) Biomarkers in assessing residential insecticide exposures during pregnancy and effects on fetal growth. Toxicol Appl Pharmacol 206:246–254.

Whyatt RM, Perera FP. (1995) Application of biologic markers to studies of environmental risks in children and the developing fetus. Environ Health Perspect. 1995 Sep;103 (Suppl 6):105-110.

Young JG, et al. (2005) Association between in utero organophosphate pesticide exposure and abnormal reflexes in neonates. Neurotoxicology 26:199-209.

Zhang X, Driver JH, Li Y, Ross JH, Krieger RI. Dialkylphosphates (2008) (DAPs) in fruits and vegetables may confound biomonitoring in organophosphorus insecticide exposure and risk assessment. J Agric Food Chem. 2008;56(22):10638-0645CrossRefMedlineWeb of Science30.